Beast or Angel?

Also by René Dubos

SO HUMAN AN ANIMAL

A GOD WITHIN

N'estez vous asceuré de vostre vouloir? Le poinct principal y gist: tout le reste est fortuit et dependent des fatales dispositions du Ciel.

Are you not assured within yourself of what you have a mind to? The chief and main point of the whole matter lieth there: all the rest is merely casual, and totally dependeth upon the fatal disposition of the heavens.

François Rabelais

Gargantua and Pantagruel, Book 3, Chapter 5 (Translation from Great Books of The Western World edition)

RENÉ DUBOS

Beast or Angel?

Choices That Make Us Human

CHARLES SCRIBNER'S SONS / NEW YORK

The author acknowledges with thanks permission to quote as follows:

On page 77, the lines from "The City" by C. P. Cavafy, translated by
Edmund Keeley and Philip Sherrard, from *Collected Poems of C. P. Cavafy*,
The Hogarth Press, and *Four Greek Poets*, Penguin Books,
by permission of The Hogarth Press and Deborah Rogers Ltd.
On page 79, the title of the song "How Ya Gonna Keep Em Down on the Farm
After They've Seen Paree," by Walter Donaldson, Sam Lewis, and Joe Young,
copyright © renewed 1946 Warock Corporation & Mills Music, Inc.,
by permission of Warock Corporation.

Library of Congress Cataloging in Publication Data

Dubos, René Jules.
   Beast or angel?

    Translation of Choisir d'etre humain.
    Includes bibliographical references.
    1. Social evolution.   2. Human evolution.   I.
  I. Title.
GN320.D8213   573.2   74-10737
ISBN 0-684-13901-4 (cloth)
ISBN 0-684-14436-0 (paper) V/P

1 3 5 7 9 11 13 15 17 19 V/P 20 18 16 14 12 10 8 6 4 2

3 5 7 9 11 13 15 17 19 V/C 20 18 16 14 12 10 8 6 4 2

Printed in the United States of America

Contents

Contents

Foreword: Conveying French in English

This book was initially written in French, at the suggestion of Monsieur A. Blanchard, director of the Paris publishing firm Gallimard Denoël. Except for letters to my family and a few speeches, I have never written in my native language since I left Europe in 1924 at the age of twenty-three. Several of my earlier books have been translated from English into French, but not by me, and I have not read the translations.

To my great surprise, writing in French proved almost effortless, and I have been told by Monsieur Blanchard that except for a few Anglicisms my text was acceptable without further editing. But a greater surprise came when I found it extremely difficult to translate my own French into English. Time and time again, in fact, I could not translate and had to convey my thoughts in other forms. If large sections of *Beast or Angel?* are substantially different from the French version, *Choisir d'Etre Humain*, it is not because I wanted to update the information or add new thoughts, but because this was the only way I could convey in English what I had tried to express in French. I have naturally been puzzled by my difficulties as translator of myself and shall try here to analyze the cause.

I wrote *Choisir d'Etre Humain* in a completely American environment—in mid-Manhattan and Aspen, Colorado. Yet as soon as I began writing in French, I recaptured the mood of my late teens and early twenties in Europe, expressing opinions spontaneously as I was wont to do with people of my age group, while walking home from the Collège Chaptal in Paris or from my office on the Pincio in Rome. In this youthful mood, I stated views about people, things, and events in a subjective manner, without much concern for precise documentation. I have the impression that remembrance of the past brought back to the surface the language of my youth which is usually buried under a layer of English. French words and turns of phrase that I had not used for half a century came to my pen without my being aware that they had passed through my mind—to the extent that I hardly recognized them when I read what I had just written. They had emerged spontaneously to express the feelings and opinions that I had developed from the immense range of facts acquired over a lifetime of reading and observation.

I speak and write English with great ease, more accurately than French, but not as casually. The reason for this difference is probably that I learned English after the age of twenty, almost exclusively from books and in a scientific atmosphere—first as assistant editor at the International Institute of Agriculture in Rome, then as a graduate student in science at Rutgers University, and especially as a staff member of the Rockefeller Institute for Medical Research (now The Rockefeller University) in New York. My usage of English has been shaped by the writing of some two hundred papers (let alone several books) on scientific problems and, since 1945, by my duties as editor of the *Journal of Experimental Medicine*. My former chief, Dr. Oswald Avery, drilled into me the ideal of carefully stating factual knowledge and objective observations, with as little use of "I" as possible and with no expression of personal feelings.

One of my difficulties in translating *Choisir d'Etre Humain* into *Beast or Angel?* was to convert the subjective responses corresponding to my European youth into statements compatible with the disciplined English of my adult professional life. In addition to these difficulties of a personal nature, there were others of cultural origin. Expressions and images that were suitable in a French context were meaningless or misleading for an American reader.

In the French text I mentioned that the names of astronauts are as poorly remembered as are the names of the *académiciens*. This refers to a parlor game in France in which people are asked to remember the names of the forty so-called immortals, the members of the Académie Française, most of whom turn out not to be known even in highly educated circles. This French "in" joke is incomprehensible to Americans.

Another cultural difference appears on page 205, where I state without qualification that it is wise, and socially useful, to take life with a smile—in French, *"prendre la vie avec le sourire."* My wife, who is American-born, pointed out to me that for her this statement evoked either a Pollyanna attitude or the "stewardess syndrome"—the mechanical smile of the airline hostess—whereas I had something very different in mind. As a young man in Paris, I had learned the very popular song, *"Il faut savoir tout prendre avec le sourire* [It is wise to take everything with a smile],"* and my

* Bracketed translations of French are the author's own.

life in Rome had confirmed me in the view that the better part of wisdom is an amused tolerance of human ways. This does not mean that one should accept the world as it is, only that the wise man is aware of the mutability of things and of the fact that truth has many faces. As Figaro says in Pierre de Beaumarchais's *Le Barbier de Séville,* "*Je me hâte de rire de tout, de peur d'en pleurer* [I hasten to laugh at everything, for fear of being obliged to weep]." The short and familiar French expression, "*prendre la vie avec le sourire,*" thus required a complicated explanation in the English version.

Most disturbing, however, was the discovery that some of the French phrases which had given me the greatest satisfaction proved to be ambiguous and obscure when I tried to translate them. One, mentioned in the Introduction (page 8), is of special interest because it is a quotation from Paul Valéry, an author justly famous for the fastidious precision of his language: "*L' homme n'est pas si simple qu'il suffise de le rabaisser pour le connaître.*" My friend Paul Horgan and I have written more than a dozen translations of Valéry's phrase, without ever rendering completely its richness in meanings and in implications. It is advisedly that I use these two words in the plural, because a searching analysis of the phrase suggests to me several interpretations. The reason for the ambiguity of Valéry's phrase is that it denotes a vast body of still undigested scientific knowledge about the biological nature of man, and also parascientific feelings about the uniqueness of man. Another example is a French phrase of my own, quoted on page 193, which suffers from a similar ambiguity concerning the role of free will in the selection of the circumstances that condition human development.

Ambiguities complicate the work of the translator but they give him the satisfaction of dealing with really human problems. All animals have means of communication which are extremely precise; however, they communicate only facts—not the associations of concepts which lead to creative thought. In contrast, human language is more than communication. It is an instrument which brings to the surface and thereby puts into contact disparate elements of one's own past, and of the collective past—thus permitting concepts to become associated and organized into new entities.

I recognized the existence of two different persons in myself as I passed from French into English and vice versa, not because I selected different pieces of information but because I placed facts in a different cultural light depending upon the language I used. I also became acutely aware of the practical impossibility of writing with sufficient clarity to permit a completely faithful translation. As a compensation for this difficulty, on the other hand, I came to believe that the ambiguities of language have provided us with a wider range of choices, thereby helping us to become more human.

Beast or Angel?

Introduction

I left my native France for Italy in January 1922, shortly before my twenty-second birthday—more than half a century ago. After working for two years in Rome, I emigrated to the United States where I have since made my home. Half a century! This is more than three-quarters of a human life expectancy and it can be a significant period even in terms of world events. Half a century is twice as long as the French Revolution and the Napoleonic era put together—from the storming of the Bastille on July 14, 1789, to the Battle of Waterloo on June 18, 1815. It is longer than the interval between the first New York–Paris nonstop flight in 1927 and the first landing on the moon in 1969. The half-century between 1850 and 1900 saw the introduction of railroads, steamships, and electricity; the telephone, the wireless telegraph, and photography; antisepsis, vaccinations, anesthesia, radiography, and the armamentarium which has revolutionized the practices of public health, medicine, and surgery.

The world of things is obviously different now from what it was half a century ago. But I doubt that there have been basic changes in life itself, and by this I mean in those attitudes and activities, needs and yearnings, which are the most important for happiness and suffering, for hope and despair—in brief, for the differences between humanity and animality. I must elaborate

3

on this paradox, because it will be in my mind throughout the present book, as I try to understand what is changeable and what is permanent in humankind and in human life.

I spent all my youth in the Ile de France, first in small farming villages north of Paris until the age of thirteen, then in Paris itself. My first trip out of France was to Italy, where I had been given a very junior appointment on the editorial staff of the International Institute of Agriculture in Rome, one of the specialized agencies of the League of Nations. I may some day try to explain—to myself first—why I left France and Italy, where I had loved the land and the people, for America of which I knew little but which was the symbol of a new kind of adventure. My only purpose here is to contrast what I experienced half a century ago with what I know of the world today.

If I were to travel now from Paris to Rome and from France to America, a few hours in an airplane would replace the long days by train or steamship that were then required. On arrival both in Rome and in New York I would find thermostatically controlled accommodations that would protect me from the cold in winter and from the heat of summer nights. The travel conditions and first contacts with new places would thus be different from what they used to be and in many respects much more comfortable. But my impressions on arrival at my destination would probably be very similar to what they were half a century ago. In Rome, I would experience as I did then a glow of human warmth from the agitated, exuberant behavior of the crowds; in New York a mechanical, seemingly heartless efficiency; in Paris an atmosphere as tense as on the eve of a revolution. Wherever I travel now I recapture much the same mood that I experienced in my youth. The reader may assume that I find little difference between the past and the present because I live in a dream world and have little contact with contemporary life. But in reality, I am very much in today's world. I love crowded city streets. I rapidly learn the mannerisms and phrases of the places where I function—even the popular songs. I deal very directly with all kinds of people. I have a tendency, which often causes embarrassment, to look them straight in the eyes, to take a personal interest in what they do, to share their enthusiasms for all kinds of causes, even the most outlandish.

One might think that I fail to recognize the extent of the change which has taken place because I too have changed at the same rhythm as the rest of the world. But this does not explain why I feel so completely at home when I go back to the places where I lived during my early youth, even though they differ so profoundly from mid-Manhattan where I have spent my adult years.

I feel at home in the small villages of the Ile de France and Picardy where I grew up, even though they seem to have remained out of the stream of modern life; at home in Paris where I lived during my teen years, even though the horrible automobile traffic keeps me now from roaming the streets as I once loved to do; at home in Rome where I first gained the illusion of being a citizen of the world, even though the spirit of the Eternal City is now crassly contaminated by modern technology; at home in London where I first encountered the Anglo-Saxon world, even though miniskirts are now more in evidence than the dignified bowler hats of businessmen and the athletic gait of pedestrians which impressed me so much when I first emerged from the Latin world; at home finally in mid-Manhattan, even though the Con Edison warnings "Dig we must" endlessly complicate the traffic in the streets that I cross every day and even though the homey brownstone houses in which I used to live have been replaced by featureless high-rise towers.

In fact, I feel almost as much at home in parts of the world where civilization seems to have come to a standstill as I do in those where progress is measured by the conquest of nature and the accumulation of gadgets. I believe that the reason for my sense of familiarity with so many places, and indeed my sense of belonging in them, is that the various forms of civilization and the forward march of progress hardly influence the most fundamental yearnings of man's nature, which are also those most attractive to me. Not only in Paris, Rome, London, and Manhattan, but in all the cities and villages of the world, the most compelling desire of normal healthy people is to satisfy certain needs which have also been those of our distant ancestors, as far back as the Old Stone Age. Unconsciously, most human beings try to recapture as often as they can and even under the most dreary circumstances certain satisfactions which can be traced to the hunter-gatherer way of life during the Old Stone Age and to a few agricultural and pastoral societies before they were contaminated by imperial ambitions and by the Industrial Revolution. There is a French proverb that says: *"Dans un coin de son coeur, on a toujours vingt ans* [In a corner of one's heart, one always remains twenty years old]." And likewise humankind continues to cherish the ways of life which were those of its youth and to search for the satisfactions which it must have experienced sometime and somewhere, in the biological paradise it has lost.

Sophisticated and civilized as we may be, we have retained from our distant ancestors the ability to derive profound satisfactions from the small happenings of daily life—when we eat, drink, and love; sing, dance, and laugh; dream, tell stories, or illustrate them in pictures; participate in events where we can be at the same time author, actor, and spectator. Both as product of a small French farming village and as a mid-Manhattan dweller, I also long for these simple but fundamental satisfactions which reflect what was best in the biological and social past of humankind. These pleasures never change character; they only take a different form. And it is because of their enduring quality that I feel I could have enjoyed living in the past as much as among my contemporaries. I believe that I participate in a great collective undertaking which began long before me and will continue long after me and of which the goal is not so much to go back to the Lost Paradise as to create the New Jerusalem. We are not only caretakers of the past; we are also responsible for the construction of tomorrow.

My purpose here is to attempt to trace the origins of needs and yearnings which have always been those of humankind everywhere and always. In this search, I shall find it natural to express the same concerns and even to use the same words when speaking of the past, of the present, or of the future—the reason being that the biological and psychological characteristics of humankind have remained essentially the same for at least fifty millennia—in other words, since the time of Cro-Magnon man. These characteristics are inscribed in the genetic code of *Homo sapiens,* and I do not see any likelihood that they can be significantly modified, either by biological accident or by human intervention.

But while the genetic nature of man is remarkably

stable, its existential expressions—its phenotypes, to use the scientific jargon—have been undergoing continuous changes in the course of time and still continue to change. The human species differs radically from animal species, not in what it is, but by what it does. We are human to the extent that we live according to certain principles which have a human quality. This quality has emerged and continues to emerge from the choices that we make throughout our individual lives and that humankind has made from the beginning of its existence. To be human is, first and foremost, to be able and willing to choose among the options that are offered to the human species by the natural order of things.

In a phrase so rich and subtle in meaning that it can hardly be translated, Paul Valéry stated, *"L'homme n'est pas si simple qu'il suffise de le rabaisser pour le connaître* [Man is not so simple that it is sufficient to consider the lower aspects of his nature to understand him]." I shall try in this book to understand the human species, not only on the basis of the biological and psychological attributes it shares with animal species, but even more by identifying its choices throughout prehistory and history. It is by these choices that animality is being transmuted into the humanity which we imagine and of which we dream.

I / Stability
and Adaptability
of Humankind

1 / Old World and New World

Around 1950, a young American physician whom I had known in New York served an internship in a hospital located on the Navajo reservation in Arizona. There he met a beautiful young Indian woman. Although she had been brought up in the Navajo traditions and spoke only broken English, he fell in love with her and they were married. When I last heard of them, they had two children and had become typical members of mid-America in a typical medium-sized city. I mention this charming couple only because the very banality of the tale symbolizes one of the most romantic adventures of humankind—the reunion of the American Indians with the peoples of the rest of the world after they had been out of contact for many thousands of generations and thus had evolved separately.

All American Indians, as well as the Eskimos, are of Asiatic origin. On the basis of present information it would seem that the human occupation of the American continent began during the Old Stone Age, approximately 30,000 years ago. Small human bands crossed from Asia to America at that time over Bering Strait, which then constituted a land bridge between the two continents. These Old Stone Age people and their descendants independently followed several migration paths and managed to spread over most of North and South America in a few thousand years. Eventually, the different groups settled each in a particular region of the continent and progressively underwent adaptive differentiations which resulted in the emergence of the various Indian tribes. The local conditions of climate, soil, topography, and the availability of natural resources naturally influenced the development of the people who settled in each particular region. Physical characteristics, ways of life, and of course language thus came to differ from region to region as each Indian group developed its individual traits.[1]

In addition to the Old Stone Age immigrants from Asia, other human groups later probably reached the American continent by sea, either from Africa or from Europe—as the Scandinavian Vikings are known to have done. But even if some of these later immigrants managed to survive in America, there is no indication that they had any significant influence on the cultural evolution of the Indian tribes. Until the arrival of Christopher Columbus and then of the Spanish conquistadors, human life on the American continent seems to have evolved independently of the rest of the species. Yet the biological and psychological characteristics of the American Indians had remained so completely "human" during all these millennia that the New York physician and the Navajo girl had no difficulty in understanding and appreciating each other to the extent of uniting their lives. From the very beginning of the Spanish Conquest, in fact, there were many cases of human comprehension, in the deepest sense of the word, between American Indians and people from the rest of the world.

When Columbus first landed on the American continent, in the region which is now the Dominican Republic, he had little difficulty in establishing rapport with the local Carib populations and wrote of their appearance and ways of life in the most admiring terms. He immediately recognized that these were not creatures fundamentally different from him and his companions, but similar human beings. His recognition of the human character of the Caribs may appear so obvious today as to be of no interest, but it was a remarkable attitude when considered in the intellectual atmosphere of the time. Before Columbus's account, Europeans had had no idea that such people as American Indians existed. Indeed, a special decision of the pope was required before they were officially recognized in the Christian world as real human beings.

The Spanish conquistadors who moved into the Americas after Columbus also readily accepted the New World natives as human. They had little difficulty in understanding the various forms of Indian civilization, even though these were barbarous from the Christian point of view. Furthermore, conquistadors and Indian women soon established associations which commonly went far beyond sexual relations, to such an extent that many Indian women came to prefer the white conquerors to the men of their own race. In any case, there emerged from unions between the two races a new population of mestizos, so vigorous that it is now a dominating political force in the Mexican state.

A few weeks after landing in Mexico, Hernando Cortez was given a young Aztec woman as a slave. Her name was Malinche and she became not only his mistress but also his inseparable companion and his political ally. Malinche remained faithful to Cortez's cause, even when the tide of war seemed to turn against him. She learned Spanish and could thus act as his interpreter with the Indians and serve as an advisor even against those of her people who were his enemies. By Cortez she had a son, Don Martin, who eventually came to a sad end but before his death played a role of some significance in the colonial history of Mexico.[2]

The rapidity with which the New World natives learned to deal with the conquistadors, and to speak their language, as well as the fact that they could develop friendly or even emotional attachments to them may have been facilitated by curious similarities in religious, social, and military organization. Like the Spaniards, the Aztecs and Incas lived in a society both rural and urban, where gold and ornaments symbolized wealth and authority; they were ruled by a military and religious hierarchy dominated by an absolute sovereign whom they regarded as next to God. But while this social similarity aided understanding between the two races, its larger significance is that it denotes among the pre-Spanish Indians an approach to life which is common to the whole human species and which has presumably its origins in the remote past. The humanizing of the American Indians antedated their arrival in the Americas during the Old Stone Age.

In North America, the Europeans found types of Indian civilization different from those discovered in South America by the Spaniards. But here again the rapport was often easy, at times giving rise to political alliances between Indian tribes and the various European nations against common enemies, either European or Indian. History emphasizes the destructive conflicts between the two races, but the similarity of their individual and collective behaviors is more important.

The biological and psychological unity of humankind is so fundamental that, even in our times, populations which had remained isolated and almost without contact with the rest of the world have been able to pass rapidly from a Stone Age way of life to the most modern forms of civilization. Eskimos, for example, readily learn to operate and to service extremely complex kinds of machinery.

The remarkable adaptability of primitive people to modern ways of life can be illustrated by two examples. Ishi, a California Indian who belonged to an extremely primitive tribe, was adopted by anthropologists at the University of California in San Francisco. Ishi was then approximately forty years old and had never known any way of life except that of his Stone Age society. And yet in a few years, he acquired many American habits, for example enjoying automobile rides about the city. He learned enough English to hold a caretaker's job in the anthropology department which had given him shelter.[3] Kiki, a native Papuan whose biography I recently read, was born in 1931 and raised in the Stone Age culture of New Guinea. As a result of his contacts with the American forces during World War Two, Kiki went to medical school and became a competent pathologist. At the age of thirty-seven, he became a political leader in New Guinea.[4]

The ease with which both Ishi and Kiki adapted to Western ways of life and learned to use the English language provides convincing evidence that the fundamental structures of the human brain are both ancient and stable. They became firmly established even before the earliest people crossed from Asia into America. As a consequence, the ways of life that developed in the New World were fundamentally similar to those of the Old World.

2 / The Saga of the Human Species

Many different parts of the world claim to be the cradle of humankind. Such conflicting claims exist because, scientifically speaking, there is no precise point at which humanity can be demarcated from animality. If the criteria by which one defines the human species are biological, then to be human simply means to be anatomically and physiologically similar to modern man. This point of view is based on orthodox evolutionary mechanisms of the Darwinian kind which are well understood and completely deterministic. But if the criteria of humanness include behavior and culture, then to be human implies social relationships, nonutilitarian activities, interest in the past, and concern for the future. Most of these criteria involve deliberate choices in the conduct of life—choices which often appear to transcend the biological laws of evolution. The generally held view is that there is an unbroken continuum between the biological aspects of human evolution and the social aspects of humanness. It is precisely this lack of discontinuity which makes it impossible to define a point at which humankind can be sharply differentiated from animalkind. The differentiation between man and beast obviously presents no problem today but becomes increasingly difficult as one looks further back into the evolutionary past.

Primates with a skeleton fairly similar to ours existed in East Africa some 3 million years ago; their brains were very much smaller than ours. It seems plausible that the genus *Homo*, of which our own species, *Homo sapiens sapiens*, is now the only representative on earth, evolved from these African primates. One line of evidence for this evolutionary origin is a series of skeletons exhibiting intermediary forms with the proper time sequence between the African primates and the modern human species, which took its present form more than 50,000 years ago. Another line of evidence is that a series of stone and other artifacts with an increasing complexity and sophistication of workmanship has been found associated with skeletons of more and more human character. As judged from present evidence, there is no discontinuity—either anatomic or in tool-making—between apekind and humankind.

Granted this evolutionary continuum, there remains the problem of accounting for those aspects of the human saga which do not have their counterpart in animal life. It can be assumed that the ape precursors of the genus *Homo* were biologically adapted to the natural conditions of East Africa where they lived and could have prospered biologically in this environment as did other animals which evolved in it and stayed in it. Even today the residents of Nairobi, Kenya, for example, and the tourists from all continents who visit the region are in accord that this part of the world provides a wonderful sense of physical comfort and exhilaration. Biologically, therefore, human beings are still adapted to life in the subtropical environment of the African savanna. But the puzzling fact is that some of the African precursors of *Homo* moved away from this terrestrial paradise and eventually settled in parts of the world to which they had no anatomical or physiological qualifications. Other animal species also moved away from East Africa, but they progressively achieved biological adaptation to their new environments by undergoing anatomical and physiological changes. In contrast, the genus *Homo* underwent only very minor biological changes. It created civilization by purely social processes, which enabled it to function and prosper in an immense variety of environments ill suited to its biological welfare. We do not know why our precursors abandoned the comfort of animal life in the environment to which they were biologically adapted or how they shifted from biological evolution to social adaptation, but we can retrace some of the steps that led them from biological safety into the unpredictable social adventures of humanity.

In the course of their early life in Africa, the precursors and eventually the members of the genus *Homo* created a tool industry of increasing sophistication. At first the tools consisted of quartz or flint fragments, crudely shaped to bring out a cutting edge. The early representatives of our genus, even though still different from us in anatomy, and especially in the size of their brains, knew how to build artificial shelters and also how to hunt big game—an enterprise which implies a high order of social organization.

Through sociocultural know-how the members of the genus *Homo* known as *Homo erectus* were able to spread as far as the southern tip of Africa, throughout Asia (where anthropologists describe them as Java man and Peking man), and into Europe, including Scandinavia. By that time, *Homo erectus* had learned to control and use fire—probably the most important single step in social evolution and in human history. The oldest hearths so far recognized seem to go back 750,000 years; they were discovered near Nice in southern France. Fire was certainly used extensively in China at least 500,000 years ago.

Excavations at the great cave of Choukoutien, near Peking, have provided a wealth of information concerning the life of *Homo erectus*. The cave is typical of the kind of spot in which prehistoric humankind chose to settle. It had ready access to water, as a river flowed by the foot of the cliff, and it provided a commanding view of a grassy plain where herds of grazing animals could be spotted. Fire played a multifarious role in the Choukoutien cave. It kept at bay even the most savage carnivores, such as cave bears, saber-toothed cats, and giant hyenas; it increased the range and value of food by improving its tastes and making it more digestible; it helped in the manufacture of tools and weapons. The ancient Greek myth of Prometheus, the hero who stole fire from Zeus, symbolizes the recognition by early people that fire had played an essential role in the emergence of civilized life.

As he moved over the world and changed his ways of life, *Homo erectus* underwent biological modifications, such as the progressive increase in the size of the brain, and social modifications, such as the increasing complexity of artifacts. But there is no particular point—no rubicon—in this stepwise biological or social evolution which can be identified as corresponding to the appearance of the first representatives of modern humankind. Paradoxically, if there is such a point it occurred during one of the Ice Ages, a traumatic period in the human saga.

The climate became much colder during the period

known as the Würm glaciation, which caused glaciers to spread over much of the earth. From the purely biological point of view, this Ice Age was completely unsuited to the human species, which was still adapted to life in a semitropical environment. The Ice Age nevertheless saw the proliferation of the Neanderthal people, who were so much like us that they are now referred to as *Homo sapiens neanderthalis*, a name which acknowledges that they can hardly be differentiated from modern humankind. The Neanderthal people 100,000 years ago had a cranial capacity as large as modern man's. They had developed a fairly complex social structure and some of their behavior patterns, which are described later, implied a high level of humanness. Their kit of stone instruments was widely diversified; in addition to spears and double-edged hand axes in many sizes and shapes, they used a great variety of tools for cutting, scraping, piercing, and gouging. More than sixty distinct types of such Neanderthal tools have been identified in France alone. Their technology enabled them to extend their settlements as far as the limits of the glaciers, but despite their extensive mastery over nature they seem to have disappeared from the surface of the earth some 40,000 years ago. They were replaced by another, slightly different human type, *Homo sapiens sapiens*, biologically identical to us, of which Cro-Magnon man is the best known representative.

By following the glaciers as they retreated over the earth, *Homo sapiens sapiens* extended the human domain as far as Siberia, New Guinea, the Pacific Islands, Australia, Japan. He took advantage of another transient glacial period which lowered the level of the oceans to cross Bering Strait from Asia into America. Ten thousand years ago, humankind thus had settled in all the regions of the earth, with the exception of the polar ice caps and a few isolated islands.

The sudden disappearance of the Neanderthal people is one of the puzzles of prehistory. A possible explanation is that the latecomer *Homo sapiens sapiens* was able to eliminate them because he had more effective weapons and better means of communication. Another hypothesis is that the two races were so closely related that they could interbreed and that the *sapiens sapiens* genotype became dominant. Both races exhibited a wide range of variability and it is possible that they formed a biological continuum with a large number of intermediary forms. This view is supported by the recent discovery in the Qafzeh cave near Nazareth in Israel of numerous skeletons which are unquestionably of the *sapiens sapiens* type but with a number of somewhat earlier characteristics. These skeletons clearly belong to the Neanderthal period and furthermore are associated in the Qafzeh cave with a type of stone industry characteristic of the Neanderthal people. The easiest interpretation of these findings is that *Homo sapiens sapiens* was closely related to *Homo sapiens neanderthalis*, having either evolved from it or interbred with it to produce the present forms of humankind. Modern man can legitimately claim Cro-Magnon man as his most direct ancestor, but he probably derives part of his genetic behavioral endowment from the Neanderthal people, even though some aspects of their anatomical appearance may appear unattractive today.

Taken together, the known facts of prehistory suggest that several branches of the genus *Homo* have evolved more or less independently in different regions of the world. This hypothesis of polycentrism does not imply, however, that the various human races are different in their genetic origin. They all belong to the same species, which probably acquired its genetic identity more than 100,000 years ago. From the genetic pool which is common to all humankind, there has progressively emerged by countless cross-breedings in the course of the millennia an immense number of subtypes which differ somewhat but are closely related to one another. This wide spectrum of subtypes within the fundamental unity of humankind has facilitated the emergence of a rich biological diversity and of an even greater social diversity.

The understanding of humanness will not be complete until more is known of its origins, because practically all manifestations of modern life are still conditioned by the prehistoric past. But what is known is sufficient to point to a paradox which accounts in part for the uniqueness of humankind: its biological stability and unity have been compatible with the changeability and marvelous diversity of its civilizations.

3 / The Races of Man

Lions and tigers may be the most powerful animals on earth, but despite their strength they do not move far away from the areas in which they were born; nor do polar bears invade the tropical forest. In nature, animals are essentially prisoners of the type of environment in which they have evolved and to which they are adapted. But the human species, as has been noted, has spread over the entire globe. One reason for the ease with which human beings change habitats is that they exhibit a rather low degree of biological specialization. They can walk, run, creep, climb, and swim. They can live on an exclusively carnivorous diet, an exclusively herbivorous one, or almost any kind of mixed alimentation. They can function and proliferate amid the sands of the Sahara or the fogs of Iceland, in the tropical forest or on the polar ice, at high altitudes in the Andes or below sea level around the Dead Sea.

This low level of biological specialization has made it easier for human beings to occupy a wide range of habitat, but it is chiefly through adaptations of a social nature that they have succeeded in colonizing the globe. For more than a hundred millennia they have transformed their environments to meet their own needs and desires, instead of adapting themselves physiologically to the conditions they find in nature. On the other hand, people who have lived continuously for several gener-

ations in a given part of the globe do progressively undergo minor anatomic and physiologic changes which
make them better adapted to their particular environment. Such biological adaptations are responsible
for the emergence of the different racial types.

To discuss or even mention the existence of racial
differences is at present almost taboo. Yet even during
the Old Stone Age there existed several races of *Homo
sapiens sapiens* which were sufficiently different anatomically that anthropologists now characterize them by
different names—Cro-Magnon man, Chancelade man,
Grimaldi man. When applied to the human species, however, the word race refers only to minor differences
which do not affect the fundamental genetic endowment.
There is no evidence, furthermore, that any of the innumerable crossings of races which have occurred in the
past and continue to occur now has brought about a loss
of genetic quality in the human species. The opposite is
more likely to be true.

The mutations which have brought about the different races have never been sufficiently profound to pre-
vent any human group from adapting to any of the
environmental conditions under which other human
beings find it possible to live. Black people originating in
the equatorial regions of Africa function successfully in
northern cities. Blue-eyed white-skinned Scandinavians
readily adapt to hot arid countries. Children of Australian aborigines who hunted kangaroos a generation ago
now live and prosper in Sydney. American Indians have
moved from the emptiness of the Far West to the
crowded cities of the East where they work on
skyscrapers in company with black, white, and yellow
people.

Stability and Adaptability of Humankind

While the genetic endowment of the human species has not changed significantly for many thousands of generations, this uniformity and stability is not absolute. As is the case for all living species, mutations continuously occur in the human species, and those mutations which have been selected by the environmental conditions and by the ways of life have produced the diversity of biological types which we call races.

Until 10,000 years ago, all human beings derived their subsistence from hunting, fishing, and gathering wild plants. This primitive way of life has certainly determined once and for all many aspects of man's genetic nature and of his behavior. But it is also true that since that time most human beings have lived as cultivators or herdsmen—occupations which require qualities different from those required by the hunting-gathering way of life and which must therefore have favored the selection of somewhat different behavioral and physical attributes. The world population remained very small—of the order of 10 million—during the immense period of time which preceded the advent of agriculture. It increased rapidly after the agricultural revolution, in part because food became more abundant, and also because most kinds of crops could be preserved more readily than could game, fish, or wild fruit, thus assuring a more dependable food supply during a large part of the year. The world population was close to a billion in the eighteenth century, before the beginning of the Industrial Revolution and at a time when the immense majority of people lived in farming villages.

It has been estimated that, all in all, some 70 billion to 100 billion human beings have lived on earth since the Old Stone Age. In view of the fact that their numbers were very small during most of prehistory, a large percentage of the total human experience must have been derived from the practices of agriculture or pastoralism. It would be surprising if these two types of occupation had not affected some aspects of human nature. Recent studies in laboratory animals and in human populations have revealed in fact that a few generations are sufficient to bring about changes in the genetic endowment which, though minor, nevertheless can have significant effects on behavior and well-being. Laboratory experiments with fruit flies and with rodents give an idea of the numbers of generations required to bring about slight but definite genetic changes in these animals.[5]

By manipulating the environmental conditions under which fruit flies (*Drosophila*) are raised in the laboratory, it is possible progressively to develop in them instincts that make them move preferentially toward a source of light or away from it, upward or downward. Such behavioral changes are inscribed in the genetic code of the insect and can be achieved in as short a time as a dozen generations. In rats and mice, such traits as aggressiveness or preference for life in plains as against life in woodlands are instincts which are genetically determined and can be altered in ten to twenty generations by manipulating the environment.

These findings acquire a special interest from the fact that, in the human species, genetic susceptibility to sickle cell anemia has been found to decrease significantly in a few generations when the environmental circumstances are favorable. The gene responsible for sickle cell anemia is extremely frequent among black people living in tropical Africa but much less so among people of the same race whose ancestors were brought as slaves to the United States centuries ago. The mechanism of this decrease in genetic susceptibility is now fairly well understood. The sickle cell gene causes an anemia, which kills many people, but it also confers resistance to another disease, namely malaria. The possession of this deleterious gene therefore has survival value in the regions where malaria is endemic, as in certain parts of Africa, whereas it presents only biological disadvantages in the regions where malaria has been eliminated, as in the United States. On the basis of these facts one can postulate that a progressive disappearance of the sickle-cell gene should occur among black people living in malaria-free countries, and one can even calculate that the decrease in incidence should become significant in less than ten generations. The study of sickle cell anemia among black people in the United States has confirmed these predictions.[6]

There are very few genetic characters—if any—for which it is possible to measure the rate of change in human populations as precisely as for sickle cell anemia. But plausible explanations of a genetic nature can be formulated for a curious phenomenon of peasant life in Europe at the beginning of the Middle Ages.

During that period European peasants were typically short and stocky, with short, almost round heads. The descriptions of the time and the figures in medieval tapestries indicate that the shape of the peasants' bodies and heads differentiated them from the feudal lords. But this description of the peasant type is not valid for all periods of European history. It applies to the serf class of the medieval period rather than to the peasants of earlier or later generations. This peculiar anatomical appearance of the medieval serf may have had a genetic cause.

Europe had experienced extensive movements of populations for many thousands of years before the Middle Ages. There were the Bronze Age migrations of people (*Volkerwanderung*) from the east and north toward the west and south; then during the Roman Empire retired legionaires were commonly settled in colonies far from their birthplaces. The innumerable crossings between different human groups resulting from these massive movements of populations may have had the genetic effects referred to as hybrid vigor. During the early Middle Ages, in contrast, the serfs were firmly attached to the land, and the feudal lords discouraged contact with the serfs of other feudal lords. As a result, the serfs had little opportunity to move out of their villages and usually married women belonging to their own demographic group. Their sedentary life in small villages and the intermarriages promoted by such a system may have been the cause of their short, stocky appearance and roundheadedness. Certainly there is evidence during that period of reduced stature and a higher percentage of roundheaded individuals in the population. The trend to consanguinity decreased with the liberation of the serfs and especially with greater ease in communication. In fact, it almost disappeared with later technological developments. The bicycle, the railroad, and finally the automobile increased enormously the range of choices of marriage partners and thus eliminated from the Western world the restricted range of matings assumed to have been responsible for the stocky, roundheaded peasant type.[7]

One can take it for granted that now, as in the past, changes in the ways of life and in the environment are progressively generating slight genetic modifications in the human species. Life in congested cities, from generation to generation, may progressively favor the selection of human types adapted to crowding, noise, and other artificial stimuli, and to occupations requiring mental alertness and mechanical skills rather than muscular strength. One can also anticipate that the genes responsible for certain pathological conditions, such as poor eyesight or diabetes, will accumulate as therapeutic procedures increase the life expectancy of people suffering from these disorders and enable them to have a normal number of children. But there is no reason to believe that such modifications will really change human nature, or that the techniques of so-called genetic engineering will affect the genetic structure of populations.

Humanity now constitutes a highly integrated and complex system of an immense number of genetic and social forces. Any profound change in its genetic endowment would destroy this integration and probably soon spell the end of human life. Yet it is probably fortunate that minor changes such as those which have been responsible for the emergence of the races of man will continue to occur, since genetic diversity can be a source of enrichment both for biological welfare and for the growth of civilization.

4 / Biological Freudianism

Although it is extremely difficult to alter significantly the genetic endowment of the human species, it is easy to shape the physical and psychological expressions of each individual person. Irrespective of color and size, all human beings possess a biological plasticity which makes it possible to modify their appearance, physiological functions, and ways of life without affecting their genetic constitution. For example, life at high altitudes rapidly brings about changes in the blood that facilitate the intake and utilization of oxygen; life in a very sunny climate results in tanning, which increases resistance to certain radiations; life in an environment contaminated with microbes stimulates the emergence of protective immune mechanisms; life in large urban agglomerations elicits certain types of behavior that decrease the deleterious effects of crowding. These effects are not due to genetic adaptations; they are not transmitted through the genes to succeeding generations.

The experience of everyday life shows that environmental conditions determine to a large extent the way the genes are expressed in children, adults, and the aged. People of Sicilian, Chinese, or Japanese ancestry who have settled in the United States have children who exhibit approximately the same statures as young Americans of Anglo-Saxon parentage. When they become adults, these children of Italian or Asiatic origin raised under American conditions have just as great a tendency as their Anglo-Saxon counterparts to become obese and to suffer cardiovascular diseases. In brief, national characteristics are less the expression of biological heredity than of the physical and social environment.

Working with experimental animals, it is easy to demonstrate that the size of the adult, its weight, its resistance to disease, as well as its longevity and certain aspects of its behavior, are profoundly influenced by the breeding conditions. The first phases of development, both prenatal and postnatal, are of extreme importance in this regard. Early influences similarly condition the development of the adult in the human species. As I have discussed elsewhere, there is a generalized form of conditioning, which one can call biological Freudianism, that extends to practically all manifestations of organic and psychic life.[8]

The potentialities as well as the limitations of the human species can now be fairly well understood—at least in biological, deterministic terms—from the interplay between genetic constitution and the conditions of life. On the one hand, genes carry the programs of the biochemical syntheses which give the body and the brain their structure. On the other hand, nutrition and other environmental forces affect the course of these processes. Moreover, practically all environmental impacts leave an imprint on the body and the mind. This imprint is likely to be almost irreversible, especially if it occurs early in life, even when the organism manages to compensate for the effects of early influences. Society thus shapes the body and the mind of the child through nutritional habits, hygienic practices, customs and tradition—in other words, by cultural mechanisms that operate in parallel with genetic mechanisms.

From a purely deterministic point of view, the adult can be defined in terms of both his genetic heritage and his cultural heritage. The immense adaptability of the human species is a consequence of this double heritage, which furthermore can be affected by the exercise of free will. This makes it possible to envisage a future world in which human society, despite the genetic stability of the species, can be improved by changing environmental forces and ways of life. Instead of following science fiction in imagining monstrous supermen devoid of really human attributes, humankind could shape itself through voluntary actions focused on reasonable goals. In particular it could give fuller and better expressions to its potentialities by taking advantage of its biological plasticity and its social adaptability.

5 / Social Adaptations

The immense majority of human beings have spent the very largest part of their lives outside of nature, in man-made environments. They are of course commonly surrounded by trees, flowers, and a few animals; they sun-bathe in summer, go skiing in winter, and take walks in the country when the weather is pleasant. But they expose themselves to nature in the raw only on rare occasions and under very special conditions. They enter into direct contact with undisturbed nature only when it is mild, open, and smiling—in other words, when it reproduces approximately the subtropical conditions under which the human species acquired its biological identity during the Old Stone Age.

Through a curious convention of language, the phrase "temperate zones of the world" is used to refer to those regions where during the past few centuries the human species has been the most successful as measured by economic and technologic criteria. But in fact, humankind could not survive in the temperate zone if it depended only on biological mechanisms of adaptation. Without fire, clothes, shelters, and various techniques for the preservation of food, human beings might die of cold and starvation during the winter and would certainly suffer from nutritional deficiencies during much of the year.

Ever since the Old Stone Age, and almost everywhere on earth, it is chiefly through social mechanisms that the human species has adapted itself to an immense range of climatic and topographic conditions. Its subtropical origin is reflected in the fact that, whatever the climate or season, human beings arrange their ways of life and local environments in such a manner that their bodies are protected from extreme heat or cold. Contrary to what is generally believed, it is almost as difficult for the Bedouins to become biologically adapted to the heat of the Sahara as it is for the Eskimos, and the reverse is true for polar temperatures. The Bedouins, like the Eskimos, can resist their environment only by means of stratagems which enable them to avoid direct contact with it—such as using the proper kind of clothing and shelter and protecting themselves against nature through other artifices.

Everywhere also, the nutritional requirements of humankind are met through social adaptations. These requirements are fundamentally the same for all human populations—whether these consume largely tubers, vegetables, and fruits as do certain Asian people, or almost exclusively animal products as used to be the case for the Masai of East Africa and for the Eskimos. Wherever they live, primitive populations succeed in obtaining diets which provide them with the carbohydrates, fats, amino acids, vitamins, and minerals that are essential nutrients for the human species. Populations which do not have access to meat or to dairy products nevertheless can manage to obtain a full spectrum of the essential amino acids from adequate mixtures of plants, supplemented by insects if necessary; empirically, they thus give to their diet a chemical composition which satisfies their biological needs.

The technologies which enabled mankind to spread over the whole globe were on the whole very simple. They consisted essentially in the manufacture by hand of stone weapons and tools, of crude shelters, and of clothing, and most importantly of course in the different uses of fire. But these simple technologies could be rendered extremely powerful because the human species has the ability to create complex social structures and especially to visualize the future; both these attributes are probably based on the use of symbols to facilitate and enrich communications. As human life spread to parts of the world to which it was not biologically adapted, its social mechanisms of adaptation required mental processes and means of communication based on a symbolic language. Almost certainly, brain and language have evolved together as the human species was in the process of acquiring its biological identity. This would explain why brain and language have the same fundamental structure in all races and why all human beings are potentially capable of adapting themselves socially to all types of civilization.

At this point, I want to digress into a discussion of the problem of language, because it occupies such a central place in human life.

According to modern theories, all human languages have a common origin and are based on a "universal grammar" which took its form at the same time as the human brain. As far as can be judged, only the human species has a true language. Chimpanzees are capable of using symbols and perhaps even a primitive syntax, but only humans encode an enormous amount of information in the form of symbols which they can then organize and transform into new creations. Certain genes which appear to be specific for the human brain enable it to recognize meanings conveyed through the universal grammar which is at the basis of all spoken and written languages. A shared universal grammar would account for the fact that American Indians and Europeans managed so rapidly to understand each other, even though their races had parted company several hundreds of centuries earlier. Similarly, in more recent times, the natives of Australia, Tasmania, and New Guinea, who still were in a Stone Age civilization at the time they were discovered, could learn to speak European languages in one or two generations. Nothing comparable has been achieved with chimpanzees, despite strenuous efforts, even when these animals have been raised in a human household from the time of their birth.[9]

In spite of the fact that the universal grammar is probably the expression of a code genetically inscribed in the human brain, the languages it engenders are not rigidly determined. They possess an ambiguity which differentiates them qualitatively from the purely biological systems of communication used by animals, and this very ambiguity may account in part for the creativeness of the human mind. Bees and other social insects have extremely precise mechanisms of communication, but this very precision probably limits the adaptability of their behavior, and therefore their creativity. Mammals too are handicapped by the rigid and narrow meaning of the signs which they use to communicate their responses to stimuli. In human languages, by contrast, some of the most useful words have such a vague meaning that they allow thought to wander freely, to roam, so to speak, and thus have chance encounters with other fragments of information. The ambiguity of language might thus be a consequence of the diversity of information that humans can recognize, register, and assimilate, as well as a source of adaptability and creativity.

The statement that man's nature has remained essentially the same since the Stone Age does not imply therefore a static view of social creations. A few facts concerning the human adventure on the American continent illustrate that the fundamental biological unity of humankind is compatible with an immense diversity of ways of life.

The human bands which came to America during the Old Stone Age were probably rather small. Yet within a few millennia they created a multiplicity of very different civilizations: in South and Central America, those of the Mayas, the Incas, the Olmecs, and the Aztecs and, in North America, those of the East Coast and Central Plains, of the Far West, and of the Pacific Northwest. Moreover, "local" styles exist in the different kinds of artifacts which have been found in the prehistoric sites. In the Southwest of France local styles have been recognized in Neanderthal stone tools 100,000 years old.[10] Even the sandals made of plant fibers some 10,000 years ago differ in shape and workmanship from one site to another within a given region.[11]

The unity and stability of the human species accounts for the continuity of its biologic development. The immense diversity of its ways of life provides the materials for a social evolution which transcends animal evolution. But in many ways the myth of Prometheus remains the deepest symbolic expression of the mysterious events which launched the uniquely human adventure. Prometheus is the symbol of those attributes which make human life different from animal life. In Aeschylus's drama, Prometheus chose to steal fire from Zeus, suffered for his audacity, and yet continued to believe that by his act of revolt he had launched humankind on the course of civilization.

II / Choosing to Be Human

1 / The Bestiality of the Human Species

The cave man is in fashion, but for the wrong reasons. His unpleasant characteristics are being publicized and used to explain modern man's misbehavior. He is assumed to have been nasty and brutish, and since we have descended from him it is claimed that we are condemned to retain the worst aspects of his nature. This would explain our propensity to kill, even to kill our fellow men; the crassness of our social relationships; our pathological desire to dominate and spoil the environments in which we live. Jean Jacques Rousseau believed that human nature was intrinsically good until it was sullied by civilization. The fashionable view at present is that human nature was bad from the very beginning and civilization has only given wider ranges of expression to its fundamental bestiality.[1]

Granted the animal origins of humankind and therefore of its instincts, social evolution brings about a progressive emergence of humanity from animality. But the use of the word bestiality to describe certain aspects of human behavior is grossly unfair to the beasts. Few animals, if any, ever engage in actions as cruel or despicable as those by which some human beings display their power. No beast has ever exploited and spoiled the earth as savagely as the human species has done through much of prehistory and all of history. The most bestial manifestations of social and individual life are found, not among beasts, but in the human species.

Consider for a moment the instinct commonly referred to as "the territorial imperative." This instinct leads many kinds of animals to identify themselves with a certain area and to fight any other animal, particularly any one of their own species, that enters this territory. Some aspects of human behavior are explained, at least superficially, on this basis. But mammals may mark their own territory by their scent and birds by singing from a tree,[2] whereas human beings use the territorial imperative as an excuse for going to war and subvert its valid biological significance by claiming political or economic rights over weaker peoples.

Since most animals live by feeding on other creatures, killing is a biological necessity. In this case again, however, the worst manifestations of bestiality are found not among the beasts, but in the human species. Recent field studies have revealed that certain animals, such as lions and hyenas, behave as if they killed for the pleasure of it, but such wanton killing seems to be unusual in the wild. Predators kill chiefly to feed themselves and rarely kill other creatures of their own species.[3] In contrast, certain human beings kill from sheer cruelty, whether it be animals, other human beings, even members of their own clan. The expression "man is a wolf for man" has remained true into the present time, although it is unfair to wolves, which among themselves usually behave as decent social creatures. Massive and useless destruction of animals by human beings has occurred in most periods of prehistory and history. Evidence of this slaughter are the immense numbers of horse skeletons which have been found at the Solutré prehistoric site in France, or the wholesale destruction of buffaloes on the American continent at the end of the last century.

The casualness with which Stone Age and modern hunters abandoned the carcasses of horses or buffaloes after the kill points to another unpleasant aspect of human behavior—wastefulness and carelessness. In fact, these traits antedate humankind since the great apes also are wasteful of food and commonly mess up their habitats. There is nothing new, therefore, in the tendency of modern people to foul their nests.

When first investigated, the cave floor of the Choukoutien cave, which had been occupied by *Homo erectus* 500,000 years ago, was littered with charred bones of horses, sheep, pigs, buffalo, and deer. More recent prehistoric sites contain food residues which had been casually abandoned by the occupants over many generations, along with artifacts of stone, bone, ivory or pottery. Such accumulations of products and objects are an essential source of documentation for the archaeologist; they enable him to picture primitive life and to study the evolution of habits and techniques. But from another point of view the materials abandoned in prehistoric sites can be regarded as the garbage of primitive humankind. They are the equivalents of the beer cans, plastic junk, radios, bedsteads, and automobile carcasses that litter modern highways and settlements.

We tend to believe that the accumulation of solid wastes is a phenomenon peculiar to our times, occurring because we are more wasteful and sloppy than our ancestors and do not have their social conscience. But the archaeological evidence shows that the cave man also was wasteful and negligent; his failings in this regard have persisted throughout history. Humankind has been careful of its resources only in periods of scarcity. The European forests looked neat until recent times chiefly because dead wood was valuable fuel; villages remained essentially free of litter only as long as objects were transmitted from generation to generation and had to be used over and over again. Only a century ago, pigs still roamed Broadway in New York City feeding on garbage. In the past, pure necessity imposed certain practices of recycling, but during eras of abundance humankind needs social discipline to enforce such practices.

Each civilization has its own kind of wastes and its own ways of being negligent. The tendency remains much the same, even though we now sweep it all into the scientific term pollution. Like the tendency to kill, the tendency to waste and to foul the nest seems to be inscribed in the genetic code of the human species.

2 / The Humanness of the Human Species

I have emphasized certain unpleasant aspects of human behavior to show that my confidence in the future of our species is not due to ignorance of its failings. This confidence is based on two different but related sets of facts. First, the human species has exhibited for at least 100,000 years certain traits which are uniquely and pleasantly human and which are more interesting than those that account for its bestiality. Second, the human species has the power to choose among the conflicting traits which constitute its complex nature, and it has made the right choices often enough to have kept civilization so far on a forward and upward course. The unique place of our species in the order of things is determined not by its animality but by its humanity.

In view of the fact that human beings evolved as hunters, it is not surprising that they have inherited a biological propensity to kill, as have all animal predators. But it is remarkable that a very large percentage of human beings find killing an extremely distasteful and painful experience. Despite the most subtle forms of propaganda, it is difficult to convince them that war is desirable. In contrast, altruism has long been practiced, often going so far as self-sacrifice. Altruism certainly has deep roots in man's biological past for the simple reason that it presents advantages for the survival of the group. However, the really human aspect of altruism is not its biological origin or its evolutionary advantages but rather the fact that humankind has now made it a virtue regardless of practical advantages or disadvantages. Since earliest recorded history, altruism has become one of the absolute values by which humanity transcends animality.

The existence of altruism has been recognized as far back as Neanderthal times, among the very first people who can be regarded as truly human. In the Shanidar cave of Iraq, for example, there was found a skeleton of a Neanderthalian adult male, dating from approximately 50,000 years ago. He had probably been blind and one of his arms had been amputated above the elbow early in life: he had been killed by a collapse of the cave wall. As he was forty years old at the time of his death and must have been incapable of fending for himself during much of his lifetime, it seems reasonable to assume that he had been cared for by the members of his clan.[4] Several similar cases which could be interpreted as examples of "charity" have been recognized in other prehistoric sites. In fact, one of the first Neanderthalian skeletons to be discovered in Europe was that of a man approximately fifty years old who had suffered from extensive arthritis. His disease was so severe that he must have been unable to hunt or to engage in other strenuous activities. He, also, must therefore have depended for his survival upon the care of his clan.[5]

Many prehistoric finds suggest attitudes of affection. A Stone Age tomb contains the body of a woman holding a young child in her arms. Caves in North America that were occupied some 9,000 years ago have yielded numerous sandals of different sizes; those of children's sizes are lined with rabbit fur, as if to express a special kind of

loving care for the youngest members of the community.[6]

Whether or not the words altruism and love had
equivalents in the languages of the Stone Age, the social
attitudes which they denote existed. The fact that the
philosophy of nonviolence was clearly formulated at the
time of Jesus and Buddha suggests that it had developed
at a much earlier date. The Golden Rule, "Do unto
others as you would have them do unto you," exists in
all religious doctrines, even in those which have reached
us through the very first written documents. It must
therefore have an extremely ancient origin.

According to the teachings of the Persian prophet
Zarathustra more than 2,500 years ago, "That nature only
is good which shall not do unto another whatever is not
good unto its own self." But since, still according to
Zarathustra, the human soul is constantly a battlefield
between beneficent and malevolent spirits, human nature
is good only to the extent that it *represses* part of itself so
as not to do harm to another person. Such an early belief
in the need to repress the evil in human nature is a star-
tling expression of the philosophy that being human
implies choices. The uniqueness of humankind comes
indeed from its potential ability to escape from the tyranny of its biological heritage. Instead of being slaves to
their genes and hormones, as animals are, human beings
have the kind of freedom which comes from possessing
free will and moral judgment. We can repress if we will
the bad aspects of our animality.

Human beings can choose not only between good
and evil, but also between other opposite traits of their
nature. For example, they can follow their natural tendency to wastefulness and negligence or their desire for

order and form. Adorning the body during life with pigments, feathers, and trinkets seems to have been a very ancient practice. What is more impressive, however, is that in the same Shanidar cave which yielded the blind Neanderthalian with the amputated arm, several bodies had been buried on beds of flowers and soft branches. At least eight different kinds of flowers, including relatives of such modern plants as hollyhock, grape hyacinth, and bachelor's buttons, have been found with the bodies. Since the plants could not have grown in the cave or have been carried by animals, the corpses must have been buried with wild flowers gathered from the hillside.[7] Other Neanderthalian bodies found in southwestern France had been painted with ocher and adorned before burial with collars and arm rings made of shells. The Neanderthalian people therefore confronted death ritualistically and with concern.

The most spectacular evidence of early man's sense of order and form is of course Stone Age art. This includes much more than the statues known as Paleolithic Venuses and the famous paintings on the walls of caves. As far back as 100,000 years ago, many tools and weapons exhibit a quality of workmanship which transcends utilitarian needs. Many ordinary objects of stone, bone, or ivory are adorned with carvings of such exquisite delicacy that they can hardly be seen without a microscope.[8] We do not know what Stone Age people tried to express through this workmanship which we now regard as artistic, but it may not have been very different from what motivates some primitive people of the present time. The Australian aborigines have stated that their nonutilitarian activities can have many different purposes, such as performing rites of magic, commemorating a notable event, or simply doing something pleasurable. The search for beauty unrelated to practical ends seems to have been a fundamental urge and may thus be an absolute human value.

Stone Age people learned to manipulate their environment so as to better adapt it to their needs. But astounding as these practical achievements are, they are trivial in comparison with the abstract concepts developed during the same period. The practice of burying the dead and the presence of objects, food, and flowers at the burial sites imply abstract notions concerning life and death. The representation of animals, plants, and human beings by various artistic techniques implies an objectivation of the external world which was a decisive step toward the view that humankind is in some way outside nature. Old Stone Age people also learned very early to make abstract notations relating the phases of the moon to the development of plants and to the migrations of animals. Indeed, many other systems of symbols have been recognized in carvings on objects 20,000 years old.[9]

These symbolic representations of natural objects and phenomena were probably at first related to such practical ends as hunting, fishing, and the gathering of wild plants. Eventually they probably facilitated the development of technologies such as those of agriculture. More interestingly, they must have led step by step to such more abstract concepts as belief in the existence of order and continuity in the universe. Concepts of this kind may be at the origin of the astonishing body of knowledge that is embodied in the construction of the megalithic monuments of Stonehenge in England and Carnac in Brittany, which antedate the Egyptian pyramids.

The uniquely human aspect of these primitive activities lies in the progressive evolution from practical and empirical knowledge into a system of scientific and philosophic abstractions. The emergence of humanity from animality consists precisely in this transmutation of utilitarian needs into an adventure of the spirit—an adventure which was not essential for the survival of humankind as an animal species, but one in which human beings have continuously and consciously been engaged since the Stone Age.

3 / Individualism and Collectivity

Humanness involves two different but complementary social attitudes. To be human implies cultivating one's individuality as fully as possible, but it also means belonging to a collectivity and therefore accepting duties and behavioral constraints which limit individualistic expressions.

Western societies encourage—at least in theory—the development of individualism. We so cherish our individuality that most people experience irritation when their names are misspelled or mispronounced, as if this represented an affront to their uniqueness. Yet it is probable that the sense of uniqueness and of self-awareness did not fully emerge until late in the evolution of the human species. Even the most independent among modern people still feel the need to belong to a social group. This is the deep meaning of Aristotle's famous statement that "man is a political animal." This fundamental need to belong has its origins in the very distant past, probably at a time when prehuman societies were still based on family bands. Eventually, the concept of belonging extended to the clan and then to the tribe.

The meaning of the word tribe is rather vague. It is used here to denote a fairly small social group, made up of populations derived from a few clans or villages located in the same area. Such populations can be assumed to have the same origin, to speak related languages, and to engage in frequent contacts. The social unity of the tribe is determined in part by similarities in the ways of life of its members and in the use they make of local resources; it also results from shared assumptions about the origins of the world and about the goals and meaning of human life. The commonality of concepts in turn constitutes a basis for the emergence and maintenance of similar religious practices and social structures. The biological continuity of the tribe and the tacit acceptance of a certain set of beliefs create a sense of solidarity among its members. But solidarity does not imply that the various members of the tribe are linked to one another by bonds of affection; it means only that they constitute an integrated biosocial entity. The more integrated a tribe is, the better it exhibits what might be called a "collective individualism" which expresses itself by patterns of behavior and thought common to all its members.[10]

The tribe is the type of social structure which has had the longest duration—probably more than a hundred millennia. In a given region the local resources inevitably play a large role in the shaping of tribal customs. People who derive their subsistence from the hunting of caribou or other big game develop habits, traditions, legends, beliefs, and taboos different from those of people who rely on salmon fishing or on the gathering of wild plants. Likewise, profound social differences inevitably develop between herdsmen and cultivators. In brief, all social attitudes are affected by the local forces which condition the ways of life, and consequently the social structures as well as the concepts of well-being. In general, traditional beliefs and attitudes survive long after the disappearance of the local conditions and resources which have brought them into being. This survival has made tribal identity a powerful force of conservatism and of social cohesion throughout prehistory and history.

In a primitive environment, the decisions and activities of each individual are determined less by his personal tastes and choices than by those of the collectivity as a whole. The individual constitutes an integral and almost anonymous component of the social body, the laws of which he tacitly accepts as an extension of himself because he identifies himself with the group. Such identification can be so complete that in certain tribes a word corresponding to "I" is at times used with the same meaning as the word "we." Even today, when a Maori speaks in the first person, he may not refer to himself but to his social group as a whole.

Humanness may owe its existence to the conditions that first caused self-awareness and the sense of uniqueness to emerge in primitive tribes. Admittedly, we may be wrong in assuming that these attributes are peculiar to the human species. How could we possibly know the feelings of an ape that continues for hours to carry its dead infant, of an elephant leaning over the body of a member of its band which has just died, or of a dog seeking the company of its master when we understand so poorly the inner life of the people who are closest to us? Every organism, animal as well as human, lives in a private world of its own to which no one else has complete access. In any case there is no doubt that human beings do have self-awareness and a sense of uniqueness. It is possible to analyze the biological and social influences that make each human being unique, unprecedented, and unrepeatable, but this analysis does not explain how and why each feels different from all other human beings.[11]

The experience of individuality depends upon a psychological separation from the external environment—in other words, upon an ability consciously to recognize as different from oneself all inanimate objects and living things, including other human beings. The Adam and Eve legend might be interpreted as the dawning in humankind of awareness that it had become different from the rest of creation. Before the Fall, humankind had lived without conflicts or worries because it was part of the natural order of things and participated unconsciously in natural phenomena. The knowledge of good and evil, however, implies an objective attitude toward the external world—both nonhuman and human. Furthermore, the sense of objectivity and therefore of separation from the rest of the world probably became sharper with the progressive increase in power to manipulate and transform the environment.

Humankind's awareness of being different from the rest of creation must date from the time, at least 100,000 years ago, when Neanderthalian people were already capable of profoundly altering their environment because they possessed an elaborate set of tools and were highly skilled in the use of fire; furthermore their complex burial practices indicate that they conceived of the existence of forces other than what they could experience by their senses. As to the Cro-Magnon people, their engravings, sculptures, and paintings make it clear that they regarded not only plants and animals but also human beings as external objects which they differentiated from their own individual persona.

The development of individuality thus begins by an objectivation of the external world, and it evolves into the realization by each individual person that he constitutes a unique specimen of the human species. The painful character of this realization is symbolized by the phrase of the fourteenth-century theologian Duns Scotus, *"Personalitas est ultimo solitudo."* Our period may have become the age of anxiety and anomie because we have developed so excessively the cult of individualism.

It is certain that the longing for individualism increases the desire, so general among human beings, to achieve some sort of isolation from their social group, at least at certain times under certain circumstances. All normal persons want to create around themselves what has been called a "bubble of personal space," a psychological space into which outsiders are not admitted, or at least in which they are not welcome. The bubble of personal space differs of course with each person and especially from one social group to another. It is much larger among Anglo-Saxon people than among Mediterranean or Latin American people. In the course of a conversation the typical North American, Scandinavian, or Englishman tries to keep himself more separated physically from the other party than does the Greek or the Arab. Latin-American men embrace when they meet—the *abrazzo*—whereas Anglo-Saxons greet each other by long-distance handshaking. The dimensions and symbolic forms of the personal space affect the quality of social relationships. Touching a shoulder, looking straight into the eyes, or bringing one's chair too close in the course of a conversation is considered as an invasion of privacy in certain cultures, whereas refusal of sensory contact is regarded in other cultures as a sign of indifference or even of hostility.[12]

Each particular social group has its own conventions for achieving separateness. In the Western world, private life goes on behind the closed doors of the apartment or of the individual house. But in certain tribes where a large shelter houses all the members of a group, some simple gestures or attitudes are sufficient to indicate that a given person wants to be isolated for a time from the rest of the community. In the seventeenth century in Amsterdam, then as now a very crowded city, René Descartes felt that he could freely pursue his meditations in solitude among the Dutch people because they were so involved in their own affairs that they ignored his presence. But in Paris in the early twentieth century Marcel Proust needed the acoustic isolation of a cork-lined room to recapture the teeming past, the *"temps perdu."*

From time immemorial, both the cult of individualism and its rejection have inspired literature and art. The coexistence of these opposite tendencies can be illustrated by two excerpts from Russian writers.

Feodor Dostoevski (1821–1881): "Man only exists for the purpose of proving to himself that he is a man and not an organ-stop! He will prove it even if it means physical suffering, even if it means turning his back on civilization."[13]

Maxim Gorki (1868–1936): "This vile life, unworthy of human reason, began on that day when the first individual tore himself away from the miraculous strength of the people, from the masses, from his mother, and frightened by his isolation and his weakness, pitied himself and grew to be a futile and evil master of petty desires, a mass which called itself 'I.' It is this same 'I' which is the worst enemy of man."[14]

True to form, the countercultures of our own times have extolled simultaneously both the most extreme egocentrism and the values of tribal life. Norman Mailer considered that the ideal of the "hip" was "to exist without roots, to set out on that uncharted journey into the rebellious imperatives of the self."[15] The hip considers himself at the center of the world and rejects the thought that his personal life is the expression of biochemical mechanisms or is governed by anonymous social forces. But in addition to the rewards of the ego trip, other countercultures emphasize the psychological comfort of belonging to a tribe. The merits of life in "communes" come not so much from the simplification of everyday existence as from the human warmth of sharing physical and emotional experiences.

The search for the satisfactions of tribal life has recently taken spectacular forms in such huge gatherings of young people as that at Woodstock in 1969 and Watkins Glen in 1973. On a lesser scale, similar occurrences have taken place throughout history and perhaps prehistory. The Dionysiac orgies of classical Greece, the pilgrimages and outdoor fairs of the Middle Ages, the seasonal ceremonies and "chants" of American Indians have all provided the opportunity for immense gatherings which cannot be entirely explained by the natural phenomena or special occasions which were their immediate causes. Industrial and urbanized societies also seem to have a need for huge anonymous gatherings, as illustrated by the size of the crowds which celebrate New Year's Eve in Times Square or the Fourteenth of July in France. The real purpose of these celebrations may not be to mark a particular event or time of the year, but rather to experience—be it only for a few hours—the warmth of tribal life. The search for unity has persisted throughout the ages as if it were a fundamental need of humankind.

Other social practices may also have originated

during early tribal life, at a time when the sense of individualism was still poorly developed and when survival
depended upon a complete sharing of resources among
the members of the community. As first emphasized by
the French sociologist Marcel Mauss in his celebrated
"Essai sur le don: forme et raison de l'échange dans les
sociétés archaiques" (translated and published in
English under the title "The Gift: Forms and Functions of
Exchange in Archaic Societies"), most primitive societies
developed a complex system of exchanges which can be
summarized by three obligations: to give, to receive, and
to repay. The simplicity of these words does not convey
either the complexity of the social mechanisms through
which the three obligations are met or the severity of the
tribal sanctions against people who break or neglect the
rules of exchange.[16] Several recent studies indicate that
the reciprocal obligations defined in Mauss's essay continue to play an important social role in the modern
world. According to the Danish sociologist Bent Jensen,
the Greenland Eskimos used to practice a complex
system of exchanges which created what he calls
"human reciprocity" and was the essential element of
their social integration.[17] When the system of "reciprocity" became useless after the Western types of social
security had been introduced by the Danish government,
the structure of Eskimo society in Greenland dis-
integrated rapidly. The obligation to give, receive, and
repay may also affect international relations between
large countries. One of the reasons that the Marshall
Plan after the Second World War resulted in psychological frictions among the participating countries
may have been that the plan did not provide mechanisms
for a reciprocal exchange between the donor nation, the
United States, and the receiving European nations.

There was no way for the Europeans to repay and therefore to complete the system of exchange.[18]

The system of exchanges was only one among many different behavioral patterns through which primitive people achieved a high level of social integration. The cult of the hero may also have contributed to this end. In general, a given social group tries unconsciously to identify itself with real or imaginary heroes who symbolize functions, attitudes, or places the values of which are widely recognized in the community. Heroes may be warriors, artists, actors, athletes; they may represent a city, a region, or a nation; they may be known for a certain political or religious faith, or merely for a style of song or of haircut. But however noble or trivial the characteristics by which they are known or remembered, their cult contributes to social integration. In most cases, furthermore, heroes symbolize certain ideals of behavior which have survival value for the community. In principle, the true hero stands not for his own interests but for those of his group. He must not be satisfied with a conventional adaptation to the events of ordinary life but must be willing to take positions which may be extreme, or even dangerous for his own well-being. He must concern himself not so much with immediate problems as with a vision of the future. Obviously few heroes practice the virtues or attitudes they symbolize, but what matters is that all known societies have needed heroes and therefore imagined them, so as to give concrete forms to their ideals of sacrifice, bravery, truth, or adventure. Long before the Biblical teachings, humankind had sensed that it could not live by bread alone and that its aspirations transcended the necessities of everyday life.

Every human being naturally cultivates his individualism and defends the rights associated with his uniqueness. But he cannot escape the influence of the attitudes and dreams that are symbolized by the traditions, customs, and heroes of his society.

4 / Humanity and the Beast

That human behavior is based on animal behavior has been repeatedly emphasized by such authors as Konrad Lorenz, Robert Ardrey, Desmond Morris, Lionel Tiger, Robin Fox, and MacFarlane Burnet.[19] This unquestionable fact does not, however, explain the differences between humankind and animalkind or the complexity of social patterns which make human societies qualitatively different from animal societies.

Ancient people had recognized the existence of a fundamental dualism in man's nature. They knew that the equilibrium which permits humanity to survive in association with its animal instincts is exceedingly precarious. The awareness of a fundamental conflict between humanity and animality and the anguish generated by this conflict are reflected in the literatures of all people at all times. On a Sumerian clay tablet 5000 years old, a father laments the behavior of his son who is so much interested in material satisfactions that he fails to cultivate higher qualities—"he neglects his humanity."[20] Zarathustra taught 2,500 years ago that at the time of creation the twin original spirits had to choose between good and evil. One became associated with truth, justice, and life; the other with lies, destruction, injustice, and death. The stuggle between good and evil has continued ever since.

Five thousand years of history justify a certain

optimism concerning the issue of that struggle and also concerning the fate of the human species. While history is replete with examples of gratuitous killing and of wanton destruction of nature, it shows also that altruism and self-sacrifice have long been acknowledged as fundamental virtues and have been widely practiced. Human life has expressed itself in great ethical systems and in spectacular material achievements. One must be blind, ignorant, or prejudiced not to recognize that man's genetic structure enables him to be generous and creative as well as aggressive and destructive. The relative importance of bestiality and humanity in human affairs is largely determined by the human choices, decisions, and actions which influence social patterns. A painful illustration of the role of social decisions in the quality of human life can be seen in the tragedy which is now destroying the small African tribe of the Iks in northern Uganda.

When the Iks were first studied extensively a few years ago, they lived as nomadic hunters and gatherers over a large and diversified mountain territory. Their life was easy and peaceful because game, fruits, and other foods were usually abundant in their native habitat and could be had without much effort. The carefree generosity and happiness of the Ik people at that time made them appear—at least to visitors—as the epitome of Rousseau's healthy happy savage.

Unfortunately for the Iks, much of the land over which they used to roam freely has been converted by the government of Uganda into a national park. As a result, they are now forbidden by law to hunt and gather food in the valleys and are compelled to farm on poor, hillside soil, where food is difficult to raise, especially in periods of drought. Food is often in such short supply indeed that old and weak people are allowed to die of starvation. Within a few years after the dismantling of their traditional culture, the Iks have become brutish, ill-tempered, selfish, and loveless; they ignore their elders and laugh at the misery of their fellowmen. They breed without love or have little regard for their mates or their children.

Colin Turnbull, the American anthropologist who last studied the Iks, has concluded from the rapid degradation of their individual lives and of their social structure that their former pleasantness was not a spontaneous expression of their real nature but was rather an artificial, learned behavior. He goes so far as to suggest that the present brutishness of the Iks is the true manifestation of their innate traits and in fact corresponds to the real core of universal human nature. According to him, natural man is fundamentally bad, interested only in himself, without any of the graces such as affection and compassion which we identify with humanness.[21]

One can accept Turnbull's observations as accurate and still derive from them conclusions concerning human nature which are less pessimistic than those he has reached. Like all human beings, the Iks have both good and bad tendencies, but the worst aspects of their nature have been brought out by the unfortunate social interventions into their lives during the past two decades. One need not be an anthropologist or a sociologist to recognize that tragedy descended on them when they were compelled abruptly to abandon their traditional hunting ways of life without being given time to adapt to agriculture.

Beast or Angel?

This is not the first time that altruism has been put in abeyance by the crude animal instinct of self-preservation. Accounts of life in concentration camps during the wars show that, under certain conditions, malevolent human beings are more to be feared than wild beasts. But history shows also that most human societies develop a certain quality of social structure which encourages mutual aid among their members, especially in times of trial. Poor societies could not long survive if they were based on the law of the jungle, which as a matter of fact has only limited application even among animals in the jungle.

The quality of a social structure is of course conditioned by the availability of resources. But it is influenced even more by the choices which enable a group of people to arrange their personal relationships and to modify the environment for the best interest of their collective life. These choices are often almost unconscious, yet they have permitted many societies to prosper under conditions even worse than the ones in which the Iks find themselves now. An obvious example, among many others, is that of Israel. That country was made into a land of milk and honey by phenomenal achievements of irrigation and agriculture in Biblical times. Then, during 2,000 years of neglect, it became a desolate land. Now it is blooming once more under the care of human beings who not only need agriculture for their survival but also love the land itself for religious, cultural, and emotional reasons.

The lesson to be derived from the story of Israel is that individual behavior as well as social and physical environment can be shaped by human will. Whereas animals prosper best in the natural situations to which they have become adapted biologically by organic (Darwinian) evolutionary forces, human beings seem to need obstacles to develop their civilizations. If humankind were just another animal species, it would still live only in the sort of mild semitropical climate in which it evolved. But its African precursors moved into less favorable areas and became *Homo sapiens sapiens* while evolving socially in response to hostile environmental conditions. When the Neolithic agricultural people started modern civilization on its course some 10,000 years ago, they abandoned almost completely the ways of biological evolution for those of sociocultural evolution.

The Sumerians in Mesopotamia literally had to create from the valleys of the Euphrates and the Tigris the land on which they developed agriculture and built the first cities. Likewise, many early people of Central America created gardens on artificial islands where they settled. The Incas created the most advanced culture of pre-Colombian America at the foot of an enormous mountain range with a narrow coastal zone. Similarly, European civilizations emerged in a part of the world which only ten centuries ago was covered by forests and marshes and which has furthermore a rather trying climate. Except for a very few quite primitive people, humankind lives everywhere on lands which it has created out of wilderness by choices and will power.

At the beginning of the Renaissance, Giovanni Pico della Mirandola expressed the genius of humanism when he affirmed that man was given by God the latitude to remain a beast or to become an angel: "With freedom of choice and with honor, as though the maker and molder of thyself, thou mayest fashion thyself in whatever shape thou shalt prefer. Thou shalt have the power to degenerate into the lower forms of life, which are brutish. Thou shalt have the power, out of thy soul's judgment, to be reborn into the higher forms, which are divine.''[22]

The creation of humanity evoked by Pico is the history of civilization. There is no need to demonstrate yet again that the anatomy, physiology, and behavior of humankind have their basis in animal nature. What is needed is a more acute recognition and a better understanding of the fact that the human species has evolved socially by developing behavioral patterns and aspirations that transcend those of animal life. The progressive passage from instinctive reactions, which are animal in nature, to willful actions has always involved painful choices and decisions. It is through these choices and decisions that humanity progressively emerged from animality.

III / The Past
in the Present

1 / The Cave and the Horizon

Museum displays and books present a rather stereotyped picture of human life during the Old Stone Age. They commonly show small bands of half-naked, muscular, and hairy people either hunting in a savanna kind of country or butchering game animals near the entrance to a cave. Most typically the scene is located in an open landscape with gentle hills, a river or some other body of water, and a few clumps of trees which do not interfere with the view of the horizon. This type of landscape calls to mind not only the semitropical regions of East Africa where the precursors of *Homo sapiens* originated but also sites in Europe and Asia where the human species first developed social life and civilization.

The atmosphere of primitive human life can be recaptured with particular intensity in the valleys of the Vézère and the Dordogne rivers in France, which contain hundreds of sites once occupied by *Homo sapiens neanderthalis* and *Homo sapiens sapiens*. From the terrace of the museum of prehistory located in the Cro-Magnon cave at Les Eyzies, one can survey the cliffs along the Vézère with their numerous caves which were used as shelters by Neanderthal and Cro-Magnon people. The Cro-Magnon cave at Les Eyzies is so large and deep that it could have sheltered 100 to 200 persons. As seen from the interior, its opening forms a majestic frame through which can be admired a vast spectacle of earth and sky. The Cro-Magnon site thus provides two aspects of the utilization and perception of space which have been associated with human settlements since the beginning of time—an enclosed area to serve as protection and an open vista leading the eye to the horizon. These two different but complementary attitudes toward space have been separately formulated by two of the most influential architects of the twentieth century, Frank Lloyd Wright and Le Corbusier.

During the first phase of his architectural career, Frank Lloyd Wright used to affirm that dwellings should above all be considered as shelters against the elements. The design of his "prairie houses" seems to invite their dwellers to withdraw into them—in his words, "as into a cave"—for protection against cold, rain, wind, and even against light. In contrast, Le Corbusier felt that dwellings should permit a direct sensory experience of the elements. During his youth he had often hiked through the Jura mountains and had been marked by the immensity of the horizon. Throughout his professional life, he tried to integrate sunlight, clouds, stars, and winds into his creations.[1] Wright and Le Corbusier thus represent two architectural philosophies which correspond to needs that the human species acquired early during its evolution. These needs have influenced the architecture of buildings and of landscapes throughout history.

In general, prehistoric populations settled in valleys where fish and game were abundant. They used for shelter the caves in the cliffs bordering these valleys, and they learned also very early to build artificial dwellings out of branches and animal hides. Whether they used natural or artificial shelters they were never far from a steppelike kind of countryside in which they could readily observe wild game.

There is no way to prove that this early visual conditioning created in the human species a dual psychology with regard to space—on the one hand the need for an enclosed limited area which gives a sense of protection, on the other hand the need for an unobstructed view of the distant horizon. But certainly both these needs exist today. If human beings are deprived of either one of these two experiences of space, they tend to re-create it in imagination to the extent of undergoing psychological disturbances. For lack of a shelter within which they can withdraw to find either protection against the physical elements or an atmosphere suitable for meditation, they withdraw within themselves psychologically to the point of denying the very existence of the external world. For lack of the opportunity to perceive the vastness of the horizon, they create imaginary worlds where they can move freely and which become in their minds more real than reality itself.

Although primitive people first utilized such natural shelters as caves, overhanging rocks, or even hollow trees, they soon began to build artificial shelters. As far back as 30,000 years ago some of these were of impressive complexity and workmanship. Throughout ancient times, technological imagination with regard to shelters has continued to manifest itself by the diversity of habitations that one finds in the different climatic zones. For example, the Sumerians early developed a complex technology of construction and architecture. As stone was scarce in Mesopotamia, they used bricks for their important buildings. In Ur, the houses of the rich were built around an inside court, the equivalent of the modern patio, in which the shade provided some relief from the ardent Mesopotamian sun. Furthermore, the best rooms opened on the inner court rather than on the outside world. In the humid forests of the tropics, even people who are primitive according to present-day Western standards have long known how to build dwellings admirably suited to the local climate; the orientation limits exposure to the sun and assures almost constant ventilation. Igloos and other forms of snowhouses are a marvelous architectural achievement because they provide excellent protection against the cold by using snow, the only insulating material available. In the American Southwest, first the Pueblo Indians and then the early white settlers made their dwellings of thick adobe walls which protected them against both the cold at night and the intense heat during the day.

Practically all human groups have thus learned to utilize local resources and to develop architectural designs for building shelters suited to climatic conditions. As one travels from the bleak and cold regions of the north to the lands of the sun, in the United States as well as in the Old World, one can readily recognize in the preindustrial settlements how the amount of snow, of rain, and of solar radiation has influenced the design, slope, and dimensions of roofs. Porches and patios, the width and orientation of streets, even the design of parks and public monuments also reflect the climatic and topographic peculiarities of the region. This folk "architecture without architects" is the more interesting because it is not slave to a geographic determinism which would dictate a single style for each set of environmental conditions.[2] The diversity of human settlements which has been imposed by environmental constraints has been enriched by another kind of diversity resulting from human imagination. For example, human beings have learned to protect themselves against cold, or against the sun, by many different architectural formulas adapted to different social conditions.

Humankind, however, has never been satisfied with shelters designed purely for protection against the elements. Always and everywhere it has tried to gain visual access to the sky and especially to the horizon. Admittedly, certain primitive peoples—such as the pygmies of Africa and a few Indian tribes of the Amazon Valley—live in the depths of the forest. But they do not live there out of choice, only to escape the pressure of more powerful populations. In any case even the so-called forest people establish their settlements in clearings which they either find or create themselves. The fact that all plants cultivated for food belong to sun-loving species is another indication that humankind acquired its biological identity, and therefore its psychological needs, in an open type of countryside.

The interplay between human life and external space has taken spectacular forms. Even at a time when the world population was still extremely small, humankind placed its monuments and its dwellings, not necessarily in the most practical places, but rather on sites from which it could dominate the horizon. The selection of high places was not simply to provide military advantage. Generally, an ideal location seems to have been one which permitted direct visual contact with the surroundings and with the vastness of the cosmos. The Sumerians commonly placed their temples and palaces on hills—at times artificially built—40 to 50 feet above the level of the plain. The most important temples were not only placed on such elevations, they were dominated by a three- to seven-story ziggurat of which the different levels could be reached by a surrounding staircase. From the top of the ziggurats, the Sumerians could observe the motions of the sun, the moon, and other celestial bodies, thus obtaining a knowledge that was incorporated into their religion and that eventually launched astronomical science.

The important buildings of ancient civilizations are generally so located that they seem to be part of the natural landscape and derive from it a spiritual and even cosmic significance which transcends their architectural quality. In Stonehenge and Carnac, the orientation of the megalithic monuments heightens the drama of the sun or of the moon when these are observed from the stone structures at certain critical periods of the year. The Stonehenge circles and the Carnac alignments probably were also the sites of departure or arrival for processions of a religious character and thus were integrated both physically and emotionally with the surrounding countryside and with the sky. The triad of pyramids in Giza; the Parthenon, Delphi, and other monuments of classical Greece; the Tibetan monasteries overlooking valleys or deserts; the mosques and the Christian churches dominating cities and villages, all demonstrate that architectural masterpieces derive much of their dramatic character from their location in the landscape and their profile against the sky. Ancient monuments naturally had inner crypts or chambers for withdrawal. But everywhere and always human beings have longed for a perception of space which would incorporate the sun, the moon, the planets, the stars, and a sweep of landscape or seascape into the far horizon. The visual apprehension of the cosmos seems to be an essential psychological need.

It nourishes the imagination and gives larger significance to life.[3]

In lands where topography did not provide dramatic views of the horizon, the landscape architects progressively learned to manipulate nature so as to provide artificial equivalents of wide open space. Perhaps one reason the art of landscape architecture reached a high level in northwestern Europe is that this region lacks spectacular scenery. France north of the Loire valley has a rather dull topography; much of it is flat and, before human beings occupied it, was covered by forests and swamps. Little by little, wide straight *allées* through the forests were cut by the feudal lords, primarily to facilitate hunting, thus opening the horizon and giving a new visual sense of space. Looking from the Château de Versailles beyond the Grand Canal, one has the illusion that the landscape is vast and open, although the distances are actually short. Similarly, other classical French parks give the impression of an uncluttered landscape and horizon, creating *"la magie des perspectives infinies"* out of a featureless nature.

On the other side of the Channel, where nature was as unspectacular as it was in the north of France, English landscape architects also managed to create the impression of large open spaces, but by designs far different from those of their French counterparts. Instead of opening long straight avenues, they created wide lawns and grassy slopes which led the eye to streams and lakes, usually man-made. In the English parks, vision can thus roam, glide, and float over a scenery made up of land, water, and sky, amid clumps of trees and shrubbery which define the plans of the design but do not mask the horizon.

The physiological need for vast open spaces has persisted in modern human societies. In Western Europe during the nineteenth and early twentieth centuries popular imagination fed on the stories of adventure in the enormous spaces of the American Far West and on the melodies evoking the seemingly endless steppes of Central Asia. Today, the word space is associated with the prodigious distances that the astronauts cross on their way to the moon, from which they can marvel at the colorful diversity of the earth across the bleak emptiness of nonterrestrial space.

The management of space by planners and architects thus transcends the solution of problems bearing on the material necessities of life. Shelters must of course protect against cold, heat, the wind, and the rain. But they must also provide environments in which humankind can satisfy the psychological need to remain in touch with the cosmic order. In modern skyscrapers, as in the Stone Age caves, a full human life implies both withdrawal into an enclosed space and the possibility of emergence into uncluttered space giving free access to the horizon.

2 / Cities Old and New

The first cities came into being more than 5,000 years ago, and Babylon already counted some 200,000 inhabitants in the sixth century B.C. Throughout history, the percentage of the total world population living in cities has constantly increased, except during periods of great social and economic disturbances, such as the so-called Dark Ages of Europe which followed the end of the Roman Empire. This continued process of urbanization has not been a historical accident, nor has it been imposed by political power. It has resulted from the fact that city life has an appeal which, although commonly denied, seems nevertheless almost universal. Aristotle wrote of the city that "while it has come into existence for the sake of life, it exists for the *good* life."[4] And the contemporary Greek poet C.P. Cavafy speaks of the city as a natural environment of man that he cannot abandon:

> The city will follow you
> You'll wander down these very streets
> Age in these same quarters
> Among the same houses finally turn grey.[5]

Practically everywhere, in modern times, farm children tend to abandon the country for the city when they have a chance. They do so even when they cannot go beyond shanty towns, in which they feel uncomfortable and alienated but which give them a vague hope of adventure and prosperity. As for intellectuals, most of them also readily succumb to the appeal of the city even though they praise nature and bucolic life.

While urban agglomerations constantly increase in numbers and size, paradoxically more and more of their inhabitants speak of urban life with bitterness; the city is accused of ruining physical and mental health as well as morals, and of never giving the satisfactions it promises. One of the ironic aspects of American history is that, since Thomas Jefferson, practically all North Americans have affirmed that they despise urban life, whereas they have consistently abandoned farms, villages, and small towns to create larger and larger urban agglomerations. Although at all periods countless urban dwellers have returned to the country in search of a calmer and greener environment, they have rarely dissociated themselves truly from urban life. In general, they have transformed the villages where they settle into suburbs which have degenerated into second-class urban settlements. It is true that cities never satisfy people entirely, but they make it difficult for those who have lived in them to escape from the urban ways of life.

Ever since the Industrial Revolution, sociologists, novelists, and poets have presented a tragic and well-documented picture of the ordeals to which the laboring classes have been exposed in the tentacular cities. Yet the cities that are the most crowded and polluted, and the most traumatic for newcomers, are the ones that grow the fastest—as if they had the greatest appeal for all social classes. The economic wealth of large agglomerations and the opportunities of gainful employment they offer account in part for this appeal, but in part only. At least as important is the hope cities give of a life richer in unexpected experiences, especially with regard to human contacts. This hope has inspired countless stories and films and is symbolized in the title of the post-World War One song, "How Ya Gonna Keep 'Em Down on the Farm After They've Seen Paree?" The city bewitches the country lad almost as much by its crowded slums as by its restaurants, shops, theaters, and displays of luxury.

There were only four cities with more than a million inhabitants in 1850, but a hundred and fifty in 1960. And, according to futurologists, there will be at least a thousand by the end of the present century. Urbanization has certainly been accelerated by the technologic and economic forces unleashed by the Industrial Revolution, but these forces have only reinforced tendencies inherent in humankind. Urban life implies a number of attributes well expressed by the old word "civility." In his *Dictionnaire françois-latin* published in 1549, Robert Etienne defined *civilité* [civility] by the charming phrase *"qui scait bien son entregent* [who knows how to deal pleasantly with other people]." Samuel Johnson refused to introduce the word "civilization" in his dictionary because, according to him, it did not add anything to what could be expressed by the old word civility. And indeed the success of urban life and of civilization continues to depend upon the practice of virtues identified with civility.

Before discussing further the merits of cities it may be useful to consider briefly the causes for the present intense antagonism to urban life, especially in the most industrialized parts of the world.

Beast or Angel?

For several generations, it has been fashionable to contrast the healthy red-cheeked country boy with the pale sickly urban child. Novels also tend to oppose the self-control of the vigorous peasant to the nervous agitation of the factory worker or businessman. But these generalizations are without basis.

While it is true that there was a great deal of organic and mental disease in the nineteenth-century agglomerations, especially among the laboring classes, this was not a consequence of urban life per se. The wretched health of these classes at that time was largely due to the fact that the Industrial Revolution had caused a massive and sudden influx of rural people into the mushrooming cities where they found detestable living and working conditions. Rural life had not prepared them for the organic and psychic stresses they encountered in their new environments. Food was usually contaminated and nutritionally unbalanced; dwellings, factory buildings, and offices were crowded and unhealthy; alcoholism and prostitution were ubiquitous—almost the only escapes from the depressing environment. We have good reasons to be alarmed by pollution and anomie in today's cities, but the situation was far worse a century ago. Dickens's novels for London, Zola's for Paris, Dostoevski's and Gorki's for Saint Petersburg, as well as countless sociological studies, leave no doubt that the ways of life in industrial environments reached their lowest levels a few generations ago. And this was as true in the United States as in Europe.

The most acute medical problems of the nineteenth century have now been essentially solved by hygiene and preventive medicine, better housing and working conditions, greater abundance, diversity, and safety of foods.

Most of the infectious and nutritional diseases of the past have thus been brought under control. Admittedly, cardiovascular disorders, various forms of cancer, and a multiplicity of ill-defined chronic conditions are becoming more prevalent in prosperous countries. Today as in the past, perfect health is a mirage that cannot be reached, but the ailments of present-day civilization are not peculiar to urban life. Statistical studies prove that health can be just as good in the city as in the country, and indeed is often better. The average life span and general state of health are fully as favorable in New York, Chicago, San Francisco, London, Paris, Berlin, or any other prosperous metropolis as among farmers, sailors, lumberjacks, or villagers anywhere on the globe. The health disasters of the nineteenth century were not due intrinsically to urban life, but to gross social deficiencies.

Despite appearances, urban life is quite compatible with mental health. Mental disturbances are fairly common among rural people at the time they migrate into large cities but that is due to stresses caused by the sudden changes in ways of life. It has been known since Hippocrates that almost any kind of sudden change can act as a cause of disease. For example, mental disturbances are observed when rural people move to jobs located in areas new to them even though still in the country; they occur even among urban people who are compelled to move from the slums into new housing developments which provide them with better living conditions. Disturbances resulting from sudden changes are not more severe in the city than in the country but they are more readily noticed because of the greater demands of the urban environment. A person who is mentally handicapped finds it easier to go shopping at the local general store than in a New York department store. The village idiot can enjoy life on the village green but has little chance of functioning well on Broadway.[6]

Most of the health problems of urban life are of purely socioeconomic origin and therefore differ from one social group to another. Instead of considering them further, I want to turn to general aspects of city life which are determined by fundamental characteristics and needs of the human species, and to the influences of the urban environment on the quality of social relationships and on the expression of human potentialities.

3 / The Clan and the Stranger

Most of the hunter-gatherers of the Old Stone Age probably lived in small bands of some fifty persons or less, organized around the family clan. Bands living in the same region must have progressively established close relationships based on the exchange of goods and of sexual partners; these relationships led them to have similar languages and eventually brought about the formation of tribes. The dimensions of the tribes were limited by the availability of food.

Population density remained low as long as humankind derived its subsistence exclusively from hunting, fishing, and the gathering of wild plants. It may not have been much greater than one person per square mile even in regions endowed with abundant food resources. The total population of France or of Spain during the Magdalenian period of the great cave paintings did not exceed 50,000. At that time, the average tribe probably consisted of 500 persons at most. Interestingly enough, the figures of 50 per clan and 500 per tribe are applicable also to primitive isolated populations which were still at an archaic stage of life without agriculture when they were discovered during the past few decades.[7]

The shift from hunting-gathering to agriculture brought about a rapid increase in the world population and led to the multiplication of villages. The size of each village was inevitably limited by the difficulties of access to the cultivated fields and to the pastures. As far as can be judged village populations soon stabilized around 500 and remained close to that level despite changes in techniques of farming and animal husbandry. Until the Industrial Revolution, the immense majority of people throughout the world lived in agricultural villages of a few hundred inhabitants. Although the Indian subcontinent is chiefly known for its enormous cities, such as Calcutta and Delhi, the majority of its population is still located in 625,000 small villages. It was while living in tribes of hunter-gatherers and in compact villages during the millennia before and after the development of agriculture that humankind acquired certain social attitudes which still condition its behavior in modern societies.

Since human nature has been shaped by the conditions prevailing in tribes and villages, the genetic code which governs the responses of the human brain probably became adapted to social relationships involving only limited numbers of people. A few recent observations seem to favor this hypothesis.

The Amish religious sect has retained almost unchanged the traditions it had in Central Europe before emigrating to America in the seventeenth century. All decisions, even those which concern private lives, must be taken in common by a kind of Quaker "sense of the meeting." But experience shows that it becomes more and more difficult to achieve unanimity of decision and peace within the community when the population of the village increases beyond certain limits. When it exceeds approximately 500, the common practice is for some of the villagers to move away and to start elsewhere a new Amish community.[8]

The academic world also seems to be influenced by the ancient conditioning to life in small groups. Problems of discipline in primary schools can generally be managed at the personal level when the children do not exceed a few hundred and the director can know all of them, but impersonal disciplinary rules become necessary when the school has more than a thousand children. A few years ago a faculty member at the University of Pittsburgh made a survey among his students and colleagues to evaluate how many persons each one of them knew well enough to remember the first name; the answers ranged from 800 to 1,200.[9]

Many adults remember more than 1,000 persons by name. The politician James Farley boasted of knowing more than 7,000 and achieved a certain fame by this kind of memory which he had cultivated for professional reasons. But knowing the name does not necessarily mean knowing the person. More important is to recognize the voice and turns of phrase; the gestures and ways of walking; certain expressions of the lips or the eyes; the habits, opinions, and familial antecedents. These are the kinds of details that permit a real identification and also a fairly reliable expectation of what the behavior of a particular person is likely to be in a particular situation. When such criteria of personality are taken into consideration, the number of persons whom we can know is rather small—on the order of a few hundred. For those who do not move much from a stable community, these few hundred persons are centered around the family, the social group, and the places of habitation and work. For those who travel a great deal, they may be distributed all over the world. I know many persons in the United States and in France, and also a few in several other countries, but the total number does not reach a thousand.

Marshall McLuhan writes about the global village in which all of us are interconnected through the marvels of electronics. This kind of relationship, however, affects our lives only in a superficial way because it does not involve either our emotions or our senses. The I-Thou relationship rarely follows from technological communications. In fact, the global village concept is of little importance for daily life because it exists only in the form of cerebral abstractions. The electronic mechanisms through which it manifests itself give us some awareness of what goes on in Tibet, in the Congo, or in any other part of the world, but in too abstract a way to have a profound effect on our organic or mental being.

In 1973, a young female college student published a book in which she presented "a chronicle of growing old in the sixties." On her own account, the most significant influence in her life had been that before she was twenty she had spent more than 5,000 hours watching television. She had witnessed on the screen John F. Kennedy's assassination, civil rights marches, student riots, the Vietnam war, space shots, moon landings—in short, all the spectacular events of the present era. Television thus had made her knowledgeable about the contemporary world, but it had not given her a real feeling for what she had watched on the screen. Although she had grown up in a comfortable and safe environment, she felt that her childhood had been more "traumatic" than that of most other people because of "the eventlessness" of her life.[10] Apparently television watching did not make her realize that misery in the American slums or the tragedies of Vietnam had been experiences incomparably more traumatic than the emotional emptiness of her own life. She had watched television as a voyeur, not as a person really involved in the human pathos of world events. Learning about the world through news reports, talk shows, television broadcasts gives the artificial thrill that comes from the illusion of proximity to events without the necessity of being involved in them; it does not elicit an organic interaction and therefore gives at best a trivial quality to the experience of the global village.

Our genetic adaptation to life in small groups also creates limits for our personal relationships. Despite all the factual knowledge we derive from radio and television, we rarely extend our concern far beyond the few persons whom we really know, and we find it almost impossible to have full confidence in those with whom we have not had direct contacts. At heart, we are still tribal people. Being adapted to life in small groups does not imply harmonious relationships with all the members of one's group. There were certainly internal conflicts in the Paleolithic tribes and Neolithic villages, but the members of each particular tribe or village knew how to interpret the behavior of the others and could understand their responses. Similarly, in the contemporary world the members of a given group are not necessarily linked by bonds of affection, but they feel at ease among themselves even when divided by conflicts. In contrast, the stranger generates discomfort, restlessness, often contempt, and commonly fear, because his beliefs, habits, and mannerisms are not readily interpreted.

Animals commonly kill any member of their own species that has been newly introduced into their community. It is probable that killing the stranger was also common among primitive people, because this was the only way they knew to protect themselves against the threats that he represented. In the civilized world, we still try to protect ourselves against the stranger by diffidence, hostility, and ostracism. Phrases such as "dirty foreigners," "*sales étrangers*," "*porchi di stranieri*" reveal that even today and in the most civilized countries the word "stranger" still has pejorative connotations.

One of the charms of a stable neighborhood is that it provides the mental comfort associated with living among people whose beliefs, customs, and prejudices we understand, and whose behavior is not likely to shock us since it is almost predictable. Because of social instability and mobility, the large urban agglomerations of the modern world rarely provide this mental comfort. Yet the need for life in small groups is so fundamental that people try to satisfy it by creating social subunits, especially in the most heterogeneous agglomerations. For example, persons of a given ethnic group tend to congregate in a particular area, as do persons of a given profession or of similar habits. The medieval towns had their Latin quarters and special streets for butchers or cabinet makers. In the same way, certain sections of modern cities become specialized in the banking or the advertising business, in art exhibits or old bookshops, even in pornographic movie houses and massage parlors. Manhattan has block parties which are limited to a few streets with a dominant ethnic character. Neighborhood schools attempt to acquire a certain autonomy to the extent of claiming the right to manage their own budgets. The phrase "neighborhood city halls" is becoming a political slogan. Paris and London retain human quality to the extent that they have remained associations of *"quartiers"* and villages.

The ceaseless mobility of the modern world constantly brings into the urban agglomerations new groups of people who constitute the modern equivalent of the Stone Age stranger on a magnified scale. Almost inevitably the initial contacts between residents and immigrants take the form of social conflicts, which are commonly regarded as the expressions of biological and psychological differences between various human groups. In reality, however, what is involved is the ancient rejection of the stranger.

The so-called racial conflicts have been especially intense in the United States as a consequence of the enormous waves of immigration which have continuously spread over the American continent since the middle of the nineteenth century: the Irish escaping from the potato famine after 1851; the Scandinavians and Germans abandoning their countries for religious, political, or economic reasons; the Italians discouraged by poverty; the Jews for whom life had become all but impossible in the ghettos of Central and Eastern Europe; the Blacks migrating from the Southern states to the industrial cities of the Northeast and Middle West; and finally the Puerto Ricans who have come to constitute a large percentage of the New York City population. These consecutive waves of immigrants, each numbering a million people or more, have created foreign bodies in the American social system.

Because the Blacks and Puerto Ricans are colored people, it has been assumed that the social conflicts associated with their presence are conflicts of races. But even more violent conflicts took place during the preceding century when white-skinned European immigrants settled on the East Coast. The Irish, the Sicilians, the Poles, the Jews then represented the "strangers" whose habits one did not understand and whose behavior elicited contempt and fear. When I first arrived in the United States half a century ago, violent action by the police against certain ethnic groups of white immigrants was not rare and social ostracism against the Jews was still almost complete. When Alfred E. Smith ran for the presidency in 1928, his Catholic religion and Irish origin contributed to his defeat.

I do not believe that the intensity of social conflicts in the United States is an expression of a special narrowness of mind on the part of the American people. It is due rather to the fact that, throughout their history, they have experienced massive immigrations which they have not had time to assimilate. The mixture of ethnic groups has not yet been homogenized in the American melting pot. During the past few decades the sudden influx of people from Jamaica and Pakistan into England, or from North Africa into France, has rapidly resulted in violence in cities as civilized and broad-minded as London and Paris, even though these immigrations were quantitatively far less massive than those repeatedly experienced by the United States during the past two centuries.

With time, fortunately, the social integration of the stranger progressively takes place. The population of all Mediterranean cities is made up of a multiplicity of ethnic groups which on the whole live peacefully together, because the mixture occurred progressively over long periods of time. In the United States, a few decades after Al Smith's pathetic failure, a Catholic citizen of recent Irish ancestry occupied the White House; John F. Kennedy became the idol of American youth. Today many Jews and Italians occupy important places in American politics. Blacks have become mayors of several very large cities, including Los Angeles. And the list of candidates for the mayorality of New York City in 1973 included a Jew, two Italians, and a Puerto Rican (the Jew was elected). Time weakens all passions and irrational attitudes, including the fear of and contempt for the stranger.

4 / Life in the City

The expressions "city" and "urban agglomeration" are commonly used as if they were interchangeable. When a difference of meaning between them is acknowledged it refers only to a question of magnitude, the phrase "urban agglomeration" being used for a very large city or for an association of contiguous cities. But there are differences of a more qualitative nature between the two expressions, as becomes clear when one considers the problems posed by attempts to create new cities during the past few decades. The spirit of a real city has subtle qualities more difficult to understand—let alone to create at will—than the quantitative aspects of an urban agglomeration.

Planners are primarily concerned with the technological efficiency of the urban system with regard to industrial, economic, and political activities. They pay less attention to the psychological and emotional needs of city dwellers or to the relation between city life and civilization. While the technological aspects of the urban system are fairly well understood and can be manipulated, little is actually known about the influence that cities have exerted on the development of human potentialities and therefore on the emergence of civilized life. Civilizations have flourished in cities for more than 5,000 years, but they have difficulty in surviving in the huge urban agglomerations of the contemporary world.

Urban planning can be used as a generalized term because its principles are much the same everywhere and have changed little in the course of time. Some 2,000 years ago, Livy had already recognized that Rome's geographical situation had played an immense part in her economic and political success.

"Not without cause did gods and men select this place for establishing our City—with its healthful hills; its convenient river, by which crops might be floated down from the midland regions and foreign commodities brought up; its sea, near enough for use, yet not exposing us, by too great propinquity, to peril from foreign fleets; a situation in the heart of Italy—a spot, in short, of a nature uniquely adapted for the expansion of a city."[11] Livy's statement is still surprisingly applicable to modern urban agglomerations. Their success depends upon ease of internal and external contacts through the efficiency of means of communication, and these are now based on the same technologies all over the world.

Whereas all urban agglomerations exhibit technological similarity, each true city has its own brand of civilization, the distinctive characters of which are derived not only from topography and climate, but also and even more from a certain quality of human relationships. In ancient Greece Sparta and Athens symbolized two different attitudes concerning the relationships between the citizen and society. During the fourteenth century, the conservatism of Siena led to a way of life and a type of civilization somewhat different from that fostered by the more adventurous spirit of Florence. Even today the similarities of conveniences, stresses, and population structures in all the great cities of the Western world do not erase their differences in mood. When judged by the spirit of the place, New York is as different from San Francisco and Los Angeles as London is from Paris and Rome.

Granted these differences of spirit between cities, the success of city life nevertheless depends upon qualities which are universal. *A priori,* it seems reasonable to assume that a sense of safety is a fundamental need, whether in cities, towns, or villages. But, in fact, safety has rarely been a significant concern in the selection or design of human settlements. Populations which have been compelled to abandon areas devastated by floods, earthquakes, or volcanic eruptions commonly return to these areas even when warned of future dangers. The smoke escaping from the fumaroles of Vesuvius or Etna does not discourage Italian peasants from cultivating the slopes of these volcanoes. Santorini island (Thera) in the Aegean Sea was the site of a stupendous volcanic eruption 3,500 years ago and has experienced multiple other eruptions during recent times, yet populations have come back after each disaster—as they have in Martinique after the destructive eruption of Mount Pelée in 1902 had killed more than 30,000 persons.

Indifference to danger can be observed in all parts of the world at the present time. A few years ago, the inhabitants of Tristan de Cunha in the South Atlantic were compelled to abandon the island after a volcanic eruption had destroyed their village. Two years later most of them had returned, despite the hospitality given them by the English government. The young people who eventually went back to England did so, not to escape danger, but because they had acquired a taste for urban life. Vestmanneyjar, the largest fishing harbor in Iceland, was destroyed in 1972 by a volcanic eruption which lasted six months. A program for the modernization and enlargement of the harbor was formulated as soon as the volcanic activity began to subside. While in Japan late in 1973, I experienced two significant earthquakes in the region of Nogoya; the inhabitants took the matter quietly even though seismologists have predicted that a very severe earthquake is due in the very near future. Smog does not make the inhabitants of Los Angeles abandon their city. Nor do New Yorkers abandon theirs because the multiplicity of its problems gives the impression that it has become unmanageable.[12]

Young people the world over are barely conscious of the dangers to which they are exposed. I have seen small Navajo children herd sheep in the solitude of the Arizona desert. I have also seen small Taiwanese children play in the narrow streets of Taipee amidst a wild automobile traffic which appeared unconcerned with their presence. Normal children and teenagers are more interested in adventure than they are concerned with safety. And this is also true for many adults. People of all ages willingly sacrifice comfort for a chance to participate in the adventures and spectacles of city life. The movement from the country to the city is often motivated by purely economic reasons; the first cities probably started as places for the exchange of goods. Evidence for trade in salt, obsidian, soapstone, copper, lapis lazuli, iron ore can be traced back 10,000 years, even before the advent of agriculture. Then, trade made facilities for defense necessary along with structures for administration. Shrines and systems of worship were also early features of city life. But the continued success of cities depended on the excitement they provided.

During the Middle Ages cities offered protection in

time of war and against robbers, but much of their appeal was probably in the fact that they made everyday life more interesting than was possible in the country. They were the sites of the ceremonies conducted by the church, the nobility, and the various professions— ceremonies which added color and glamour to medieval life. In the Moslem world, the market and the mosque each Friday are the centers of collective life; the requirement that a minimum of forty men be present for the noonday prayer on Friday tends to associate religious practices with fairly large population centers. One hundred years ago, the large European cities were crowded and unhealthy, but their atmosphere was electrified by the hopes emerging from the Industrial Revolution. Modern cities are the stages for most of the great happenings of the present epoch; they give the impression that each person can participate in these events to the extent that he wishes.

One of the greatest contributions of cities is that they have provided mechanisms for making the presence of the stranger tolerable and for facilitating his integration into the social body. The Italian piazza, the Spanish plaza, the French mall have long played a role analogous to that of the agora in the ancient Greek cities. People of all classes and origins meet in these public places and can become acquainted without having to commit themselves to personal relationships. Throughout history, it has been in the public places that the resident has first become aware of the stranger, observing him critically but also with curiosity, becoming used to his mannerisms before engaging in nodding acquaintance with him, and finally seeking his company. In the miserable sections of Calcutta, the public places available for the activities of daily life bring about casual contacts among people belonging to different religions or social classes and thus facilitate collaboration in neighborhood activities of people who would otherwise have little if any chance of becoming acquainted.

Public places emerged spontaneously in all cities of the Old World. But although the role they have played in the past is well understood, their establishment in new cities has rarely been successful. It is easy to design attractive public places where people can assemble, but it is difficult to generate the human warmth that comes from collective activities. A real city environment depends upon an atmosphere in which the human presence is active rather then passive. Public places are elements of social integration only to the extent that they encourage and facilitate the kinds of activities identified with the words "happening" or "occurrence," events that emerge spontaneously and almost unconsciously.

The quality of the townscape therefore involves much more than natural and architectural features. People rarely visit the city or settle in it to look at parks or monuments, however beautiful these may be. They go to the city for the sake of human encounters and of all that comes from them—social contacts, business and employment opportunities, intellectual and artistic satisfactions. Only when these social expectations have been satisfied do people become really concerned with the physical amenities of the environment. In fact, it is probable that environmental amenities can be usefully defined only in terms of the contributions they make to the human encounter. People are more likely to gather where something is happening that interests them, moves them emotionally, or gets them socially involved. The popularity as meeting places of railroad stations in the past and of airports now does not come from the physical pleasantness of this kind of environment, but from the fact that it is associated with adventure through travel and human contacts. The shops of artisans or tradesmen also have been popular meeting places throughout the ages because they provide displays of human enterprise.

The situations which have the greatest appeal are the ones in which the spectator can take an active part in the act. The piazzas, malls, and open-air cafés in Europe are examples of such situations. For complex historical reasons, public places have not been popular in the United States since the end of the nineteenth century, and even the village greens of New England have lost much of their social function. But increasingly during recent years, the doorsteps of old houses and the low walls surrounding the fountains in front of new sky-scrapers are used as seats by strollers, workers, lovers, and bums even in the busiest New York thoroughfares. Watching the life of the city as it goes by provides endless entertainment for the spectator who himself contributes to the spectacle by his attitudes and his remarks.

Cities are loved, not so much for their natural and architectural splendors as for the variety and intensity of the spectacles that the ordinary events of human life generate in the streets, the malls, the piazzas, the parks, and other public places. The human, as against the economic, success of a city is measured by the opportunities it provides for its inhabitants and its visitors to participate in its collective life.

As mentioned earlier, the biological limitations of the human brain make it difficult to know really well more than a few hundred persons. A group of this magnitude therefore constitutes the most comfortable unit of social life, but in most types of human settlements this comfort is bought at a high price. Life in the tribe, the village, or the small neighborhood can offer only a narrow range of human associations, and it imposes behavioral constraints which limit personal development. In contrast, the city offers both a wider range of choices and greater freedom of action. The variety of places of work and of entertainment, as well as of such specialized groups as churches and clubs, provides a broad spectrum of activities and relationships among which the city dweller and visitor can choose. Even more important, the streets, plazas, or malls, the cafés, restaurants, and places of entertainment provide the opportunity for chance encounters which are at times extremely rewarding, precisely because they add unexpected components to life. As these encounters do not commit the participants to continued associations, they give time for deliberate choices to those who wish to escape from an accustomed social milieu. The hope that one's horizon will be enlarged by accidental contacts contributes everywhere to the appeal of city life.

Civilizations owe a great deal to the chance encounters which have enabled persons of different origins and talents to enrich themselves reciprocally and to formulate together new modes of life, of artistic expressions, and even of scientific knowledge. The interplay of thoughts in the agoras of ancient Greek cities helped to sharpen philosophical and political concepts. In Europe during the seventeenth century the search for new ways to manipulate the physical world led to the creation of academies and thus accelerated the development of science and technology. The revolt against the academic painting of the nineteenth century became an organized movement through a few artists who met in the art centers of Europe. Impressionism might not have developed into a full-fledged movement if Claude Monet, Auguste Renoir, Alfred Sisley, and Frédéric Bazille had not met in 1862, first in the stultifying art classes of the Ecole des Beaux Arts and then in the Paris cafés and restaurants.[13] The history of modern art is largely the history of chance encounters which developed into irreversible historic trends. The Impressionist school and the other schools that followed it arrived at their philosophies of art not by abstract thought but as a result of personal contacts between congenial souls. Humankind has progressively discovered its intellectual and emotional wealth through the unpredictable encounters and confrontations made possible by life in the city.

Most human beings desire to participate in the adventures and spectacles of collective life, but they also want to make a unique creation of their own life. All share the biological endowment of the human species; all retain some psychological aspects of the Paleolithic hunter, of the Neolithic farmer, and of the urban dweller who has experienced the splendors and miseries of civilizations; all are at the same time Don Quixote and Sancho Panza, Dr. Jekyll and Mr. Hyde, the Tartarin of Tarascon who dreams of hunting lions in Africa and the one who loves his cup of chocolate in bed. Yet with so much in common, each one knows that he is a unique specimen of the human species.

Genetic constitution and the accidents of life make each person different from all those who have lived before, who live now, or who will live in the future. Furthermore, the freedom to move into new environments constitutes a mechanism for self-discovery and self-realization. Each human being first imagines what he would like his life to be. Then he tries to shape it by taking advantage of the options open to him, and especially by seeking the environments that seem to him most suitable for the fulfillment of his dreams. We do not react passively with environmental factors; we actively move toward them and we respond to them creatively. The teachers of the young Napoleon Bonaparte predicted in his school records, *"Ira loin si les circonstances le favorisent.* [Will go far if circumstances favor him]."* But in a large measure, it was Bonaparte himself who selected or created the circumstances that enabled him to go far and to become known as Napoleon.

The chief merit of the city may be to provide a wider range of options to act out one's own way of life. In this respect, it is important to note that, while the child born and raised in the slums is theoretically as free as the privileged child, his range of choices and his ability to move are so limited that he becomes almost a prisoner of biological determinism. The diversity of the environment is therefore more important than its efficiency or its beauty because it provides a wider range of circumstances for individual development. Environmental diversity helps each human being to discover what he is, what he can do, and what he wants to become. The great cities of the world acquired a great diversity from their historical past and this is one of their greatest assets. It may be exhausting to live in New York, London, Paris, or Rome, but each of these cities, as well as others that readily come to mind, offers an immense range of intimate atmospheres and public spaces that provide stages on which to create one's own self-selected persona, while functioning as an organic part of a social group.

5 / Crowds and Machines

The earth as a whole is overpopulated, but this is not what makes life in modern urban agglomerations more traumatic than city life used to be. The population density in the heart of Manhattan, London, Paris, and several other great cities of the Western world was somewhat higher a century ago than it is today, but crowding as an experience is more painful now than it was then. High population density does not therefore completely explain the unpleasant experience that we call overcrowding.[14]

High population densities are as old as civilized life and perhaps as human life. While the total world population of the Old Stone Age was small, it was concentrated in a very few spots; there may have been 200 persons in the Cro-Magnon cave at Les Eyzies. Population density was high in the Neolithic villages, in the Bronze Age cities of Mesopotamia and the Indus Valley, in Imperial Rome, and also in Tenochtitlán, the pre-Spanish capital of the Aztecs. As for medieval towns enough of them have survived to give a visual impression of the way their inhabitants lived on top of one another in the small houses and narrow streets, especially when town walls restricted expansion into the countryside. In the immense Kalahari desert the primitive !Kung people live today in a few isolated settlements which are extremely crowded, entirely by choice.[15] In brief, crowding within human settlements has been the rule rather than the exception.

Humankind adapts easily to high population densities because it has experienced such conditions throughout its biological past. One important difference today, however, is that people no longer have the ready access to the countryside that until a few decades ago was available to the inhabitants of even the largest cities. Whatever the size of the human settlement little time and effort was required to escape from the crowds. Another difference is that, almost universally now, crowds imply machines. When Jean Paul Sartre wrote in *Huis Clos* [*No Exit*], *"L'enfer, c'est les autres,"* he should have written: "Hell is other people and their machines."

Human crowds per se are not responsible for the nervousness of life in the modern agglomerations. We suffer less from contact with people than from exposure to the unnatural stimuli generated by the machines that accompany them everywhere in the industrial world. Motor cars, motorcycles, telephones, radios, television sets, and other gadgets enslave us to a nonhuman and often antihuman environment. The horror of the physiological and psychological responses to excessive mechanical stimuli can be read in the expressions and behavior of human beings all over the Western world, even in situations where the numbers of people are limited—the worried distortion of faces, the brutality of contacts in the street, the indifference of passers-by who attempt to protect themselves by ignoring their environment and even the other pedestrians whose path they cross.

Behavioral patterns in crowded environments, however, usually constitute only defense mechanisms akin to putting on a superficial protective layer. I have witnessed how rapidly tenseness and indifference can be replaced even in Manhattan by smiles and congeniality when some of the avenues are closed to automobile and motorcycle traffic. The crowd of pedestrians is then much denser than usual. People of all races, all ages, and all social classes invade the pavement, but instead of the violence and incoherence that might be expected a new order emerges from the crowd. The general conversations form a background for the chattering of children and for the exchange of jokes. Human voices become orchestrated instead of being shrill as they usually must be to overcome traffic noises. And one can hear the church bells! There are beautiful bells in mid-Manhattan and if they are so little known and appreciated it is only because their ringing is drowned out by mechanical noises. Dense as the crowds may be in the car-less avenues and streets, the smile comes back on the lips and in the eyes of the strollers. It is then pleasant to remember that Manhattan is at the longitude of Naples and that large sections of its populations originated from the Mediterranean shores, Central Europe, Africa, and the Caribbean, where large crowds evoke festivity.

I recall a particular Sunday in late November with a soft gray sky and a gentle temperature—almost a feeling of no temperature at all. Such weather is not typical of New York in the fall. It had neither the exhilarating quality of a high, brilliant sky nor the gloomy effect associated with a pervasive rain. The atmosphere was just one of relaxation after the excitement of the summer. Central Park was crowded but did not appear so. More exactly, it was alive with people of all ages and all colors, who were enjoying a world temporarily free of motor cars. The great creative innovation in Manhattan during recent years has been, not the spectacular new buildings, but rather the elimination of automobile traffic from certain areas on certain days, in particular from the parks on weekends. On that day children and dogs ran about freely, chasing each other around trees and up the huge boulders. Couples, young and old, were affectionately holding hands, as in a private pleasance. All sorts of people in all kinds of colorful and pleasantly ridiculous accouterments talked, laughed, and especially smiled, with the same kind of smile that New Yorkers wear on the avenues when these are free from cars. The bicycle riders moved silently everywhere, some mildly intoxicated by the physical effort, others going slowly with an expression of beatitude on their faces. New York can be wonderfully human when automobiles are absent; it gives then the sense of relief one experiences after awakening from a nightmare.

I have dwelt on these aspects of Manhattan life which are generally ignored in order to convey my belief that the brutality and heartlessness of modern agglomerations does not come from a loss of human qualities, but rather from constant exposure to mechanical insults which paralyze the expression of generosity and friendliness. Cities can fulfill their ancient mission as civilizing forces only if they provide environments in which the human encounter becomes once more a pleasant experience and releases the potentialities and subtleties of humankind. Throughout history the human encounter, even in the midst of crowds, has been an inspiration for personal and collective fulfillment.

6 / *Hauts Lieux* and Monuments

A few places acquired very early an emotional significance which they have retained despite the vicissitudes of history. They are called in French the *hauts lieux* (high places) of civilization, a term denoting that they symbolize spiritual values which have marked the ascent of humankind. *Hauts lieux* can be mountains such as Olympus, Tabor, or Fuji; hills such as the Acropolis, the Mount of Olives, or Vézelay; bodies of water such as the Sea of Galilee, the Nile, or the Ganges; points of pilgrimage such as Santiago de Compostella, Mecca, or Trondheim; holy cities such as Jerusalem, Benares, or Rome. Few of these places are notable for material comforts, economic wealth, or political power. But they all evoke cosmic meanings and spiritual aspirations.

Ever since the Old Stone Age, humankind seems to have felt the need of expressing certain collective pre-occupations by activities which do not have any obvious practical usefulness. The Neanderthal and Cro-Magnon people shaped and decorated their tools and weapons with a fastidious care which had no bearing on practical effectiveness. Many tribes of hunter-gatherers devoted prodigious efforts to the creation of paintings in the depths of caves where they did not live. Prehistoric humankind created the stupendous megalithic structures of Northwestern Europe as well as the colossal enigmatic statues of Easter Island. The urge to create monuments having chiefly a symbolic value has continued throughout history. We owe to it the Sumerian ziggurats; the Egyptian pyramids; the temples, cathedrals, and shrines of all religions; the statues and triumphal arches dedicated to gods and heroes or to the great events of national life.

Formal ceremonies are also of very ancient origin. Their processions and banners, their accompaniment of songs, drum rolls, clarion calls, ringing of bells, and orchestral music acquire both biological and social meaning from the fact that they associate the memory of events with sensory experiences. By a kind of reflex conditioning, these associations affect the subsequent behavior of populations and thereby the shapes of societies. The roll of drums can evoke the great deeds of heroes or the military execution of the traitor. The national flag is never without significance, whether one salutes it with respect or desecrates it to express contempt or hatred. No one can entirely escape from the conditioning created by the ceremonies of a particular society.

Great ceremonies had at first a magic, religious, or civic motivation. The circles of Stonehenge and the alignments of Carnac, while they probably served as solar and lunar observatories, were also, as already mentioned, the scenes of processions which celebrated significant positions of the sun or the moon in the firmament in relation to the surrounding landscape. The Hindu festival Kumbh Mela attracts immense crowds which participate along the Ganges in ceremonies which date from the Stone Age. Most of the popular festivals of present-day Europe also originate from primitive celebrations which were adopted by Christianity before they acquired their present secular character. It is only during recent decades that these have become merely a pretext for holidays which are becoming increasingly trivial because they are no longer holy.

The great gatherings of people at the time when certain important natural phenomena occur may have had a biological source, perhaps analogous to that of the gathering of animals of a given species at certain times of the year. The seasonal migration of fishes, birds, or mammals; the assembly of large numbers of males and females of the same species at mating time, followed by their dispersal for the rest of the year, illustrate how deeply animal life is ruled by cosmic forces, often of such subtle character that they are not readily identified. Human beings do not escape the influence of these natural forces, even when their activities appear to be entirely determined by cultural attitudes. In industrial as well as in primitive societies, crowds of young people spontaneously assemble in springtime, not only because of sexual attraction but perhaps chiefly to satisfy a psychological need for togetherness.

Although the need to gather in groups expresses itself through a multiplicity of forms which have a local character, its initial motivation in the far distant past was probably biological. The hunting of large game and the gathering of wild plants among the Paleolithic hunter-gatherers and later the seasonal occupations imposed by agriculture and animal husbandry must have been among the first influences to give rise to large gatherings of people and also to ceremonies probably of magic nature. Recent excavations at Monte Alto in Guatemala have brought to light an ancient Mayan monument so located that the sun is directly over it at the time of the winter solstice; the sun also strikes two other buildings in the same group precisely at the time when the land must be prepared for corn planting. Even today, these solar events are accompanied by rites in which the whole population participates, because agricultural practices are connected with religious ceremonies.

Plowing time, sowing time, harvest time are still commonly associated with various types of celebrations: agricultural fairs have become occasions for seasonal and regional festivities. In urban or industrial areas, where the reasons for marking the rhythms of nature are no longer apparent, populations invent new pretexts to gather in large crowds which develop some kind of emotional unity despite their social heterogeneity. Our era has witnessed the immense processions and other collective manifestations which marked the ends of the two world wars; the explosive public rejoicing at the news of Lindbergh's flight across the Atlantic; the scenes of silent despair after the murder of President Kennedy. In New York City, hundreds of thousands of people who could, on their own televisions in the comfort of their homes, have seen the astronauts of Apollo II set foot on the moon elected instead to watch the event on television screens in public places; they felt the need to participate in the event with an immense crowd so as to experience in full force the collective emotion.

The need to share in a collective experience can thus be detected in prehistory and followed throughout history. Until our times, furthermore, human beings have devoted a large percentage of their efforts and resources to collective enterprises. Their most original achievements relate to preoccupations of the social group, rather than to the satisfaction of their own material needs. The vastness of the Lascaux galleries, and the knowledge that hundreds of other caves in France and in Spain also shelter monumental wall paintings of the same period, help one to appreciate the social effort represented by the Paleolithic rock art. Great technical feats were required to paint in the darkness, on almost inaccessible walls and roofs of the caves, large groups of animals the shapes of which were made to fit the irregularities of the rock. And yet these feats were accomplished at a time when the populations of France and of Spain probably did not exceed 50,000 persons. Thus local populations which would now be regarded as far too small to produce anything beyond the satisfaction of their immediate material needs devoted a significant percentage of their efforts and skills to creations which had only a symbolic significance.

A prodigious collective social effort was also required for the construction of the megalithic monuments. The huge stone blocks used in Stonehenge had to be imported to Salisbury Plain from great distances; their transportation, erection, and architectural organization must have presented enormous problems, especially in view of the small numbers of people involved. The magnitude of the social effort represented by the Carnac alignments can be evaluated by remembering that more than 4,000 megaliths were brought in from other areas and were erected according to a certain orientation over a length of several kilometers.

As during prehistory, each historical period has devoted the best of its genius to nonutilitarian activities. The creation of temples, monasteries, cathedrals, palaces, civic buildings, parks, and pleasure gardens was more often than not the feat of small communities. Athens was still small when it began to cover the Acropolis with the multiplicity of its monuments. Most of the cathedrals, palaces, and other edifices erected during the Middle Ages and the Renaissance were in cities which would now be considered very small towns. Chartres, Bourges, Rheims, Amiens, Beauvais counted only a few thousand inhabitants apiece when they built their gigantic cathedrals, and the same is true for Canterbury, Winchester, or Salisbury in England; for Assisi, Siena, or Pisa in Italy.

Most of these cities are now much larger than they were then, yet they find it difficult to maintain in good condition the architectural treasures they have inherited from the past, let alone to create new ones. It is true that the medieval monuments required several decades for their completion, but this does not affect the argument. After more than fifty years, during a period of continuously increasing prosperity, it has proven impossible to complete the construction of the cathedral of Saint John the Divine in Manhattan because the project now appears too ambitious. Thus even the richest of modern societies cannot mobilize the resources required to build or maintain the kinds of monuments that small towns have erected throughout history with technical means far inferior to those of the present. In shaping civilizations, judgments of value are far more important than material resources and technological know-how.

Modern societies can of course boast of numerous achievements which go far beyond those of preceding societies. They have improved the physical well-being of all social classes; they have created phenomenal means of transportation and of communication which are available to the poor as well as to the rich; they have built residences and industrial plants possessing a kind of comfort and efficiency unknown in the past. But this very creativity and prodigality in the practical aspects of life put in even sharper relief the poverty of modern achievements in most domains of emotional and spiritual life.

Future generations will probably admire our era for its theoretical science, its adventures in space, and its medical discoveries. They may even admire the parkways which, though of limited practical usefulness, have the merit of opening landscapes that previously were almost inaccessible. Modern science and technology therefore constitute expressions of the genius of this era which have a collective value. But it is doubtful that these expressions will equal those of the past in imaginative richness or in influence. When evaluating them, it must be remembered that for a hundred millennia human beings devoted a large percentage of their resources and inventiveness to creations which did not contribute to their material needs, but which still enrich life today.

The nonutilitarian aspects of human life are so universal and so ancient that they must correspond to a psychological necessity. *Hauts lieux* and monuments may indeed derive a truly biological value from the fact that they symbolize the memories and aspirations of the social group, thus contributing to its integration in space and in time. If this interpretation is justified, the word monument need not imply gigantic buildings, or even man-made structures. The holy rocks of the Aborigines in the Australian desert, the dancing grounds of American Indian tribes, large trees and other notable objects of the landscape become veritable monuments when they are endowed with symbolic values by the history and imagination of the tribe—and thus incarnate its pre-occupations and aspirations. Such symbolic embodiments of the tribal spirit are the equivalent of the religious altars or the tombs of the unkown soldiers which symbolize the spirit of religions and of nations. The nature of the monument is of less importance than its power to evoke collective emotions in all the members of the social group.

An efficient organization assuring that all material needs are satisfied practically and pleasantly is not sufficient to make a viable organism out of a society or a city. An additional requirement is the possession of what might be called a soul. Religious beliefs and patriotic feelings have long been the chief source of inspiration for the nonutilitarian aspects of existence, but they can no longer act as integrative forces. They must be given new meanings to express the new contents of beliefs and loyalties. As mentioned earlier, a new form of religion might emerge from a deeper awareness of our relation to the earth and to the cosmos. Patriotic feelings might be reborn from the need to achieve some form of identification with the physical and social environments.

The soul of a society or a city is a concept as vague as that of the soul of a person. And yet these expressions correspond to realities of which the existence is obvious to all sensitive persons. A historical past which has left visible traces, triumphs and tragedies which survive in the collective memory, habits and hopes which are manifested in the form of achievements or projects are all values which create the identity and uniqueness of a social group. These values are generally best expressed by concrete symbols which are apparent to the senses and meaningful to the mind—in other words, by ceremonies and monuments. Modern societies may no longer have the desire, or perhaps the vigor, to build pyramids and cathedrals. But they need the equivalent of these structures to express new aspirations—the thoughts and dreams which must replace the beliefs that have been lost. Human societies need ceremonies and monuments, not to advertise their power and glory, but to symbolize their memories and hopes. Ceremonies and monuments are the incarnations of the spirit which animates all human societies; they are the embodiment of the soul which makes cities and nations spiritual organisms by linking their past to their future.

IV / At Home on Earth

1 / Yesterday's Future Shock

There have always been timid souls who believed that adaptation to the future would be difficult, painful, and perhaps impossible. Each period has had a prophet to dramatize the problems inherent in "future shock." As Alvin Toffler puts it for our times, "In the three short decades between now and the twenty-first century, millions of ordinary psychologically normal people will face an abrupt collision with the future." They will be so "overwhelmed by change" that they will experience future shock.[1]

Toffler presents much factual information to support his thesis, and I agree with him that difficulties are bound to arise from the collision with tomorrow. But I doubt that the transformations of human life during the next three decades will be as drastic as he believes. Furthermore, I know that humankind has repeatedly survived ordeals at least as severe and as socially upsetting as the ones our societies are likely to experience before the end of the century. There is little new in future shock, except for the felicitous phrase. The best sedative in periods of political agitation and concern about technological innovations is a strong dose of history.

None of the inventions introduced during the past fifty years has had on daily life effects as profound, as sudden, and as widespread as those generated by the steam engine and electricity a century ago. Not even the airplane has revolutionized life as much as is commonly stated. The first flights of the Wright brothers at Kitty Hawk in 1904, the first crossing of the Channel by Louis Blériot in 1909, the first nonstop New York–Paris flight by Charles A. Lindbergh in 1927 naturally generated immense excitement in the general public all over the world. And it is almost incredible that, so few years after these marvelous feats, jet planes now cross continents and oceans in just a few hours. But although this technological achievement has changed business practices and made the earth look smaller, it has had only a minor effect on daily life. For most people the experience of flight is now largely one of boredom that airlines try to relieve with trivial music and films.

As mentioned in the Introduction, some material aspects of life have been profoundly transformed in the course of two half-century periods during modern times, but profound transformations also took place rapidly on several occasions in the distant past.

Fifty centuries ago, the written word had begun to replace the oral recital of possessions, history, and legends. Cities were developing in Mesopotamia, Egypt, India, and Pakistan, with the attendant problems of urban life. The gap between rich and poor was widening. Regardless of economic status, becoming an urban dweller meant unpleasant changes in habits, including learning to sit on straight uncomfortable chairs and to sleep on beds instead of finding rest wherever and whenever convenient on the soft earth or on straw as had been done in the past. Urban life was already marked by mass tragedies and personal worries. Ancient Sumerian texts reveal the deep attachment of writers to their cities—the last line of one poem affirms proudly, "I am a

Sumerian''—but many texts express anguish at the damage done to cities by the accidents of nature and by wars. They reveal also that the conflict between generations was even then a feature of prosperous life. A scribe complains bitterly about the ill-conduct and ingratitude of his son who refuses to follow his traditions: "I, night and day, am tortured because of you. Night and day you waste in pleasure. You have accumulated much wealth, have expanded far and wide, have become fat, big, broad, powerful, and puffed up . . . your kin wait expectantly for your misfortune, and will rejoice at it because you looked not to your humanity."

In Egypt also, social disturbances had followed the first period of great wealth and power. "The land spins around as a potter's wheel does! The robber is now the possessor of riches. All maid servants make free with their tongues! When their mistresses speak, it is burdensome to the servants." "Doorkeepers say: 'Let us go and plunder.' . . . The laundryman refuses to carry his load. . . . Foreigners have become people everywhere!" Families of position and wealth were being unseated by the *nouveaux riches*. The old tombs were neglected, even those of the wisest of men. Since nothing is permanent, wrote an ancient Egyptian poet who might be speaking for our own times:

. . . let your desire have play . . . as long as you live! . . .
Make holiday, and do not grow tired of it!
See, no man is allowed to take his property with him.
See, no one who departs comes back again!

Other Egyptians of the same period, however, believed that life should be governed not by materialism but by moral values. Since riches and power are not lasting, perhaps it can be hoped that moral character will endure. A king advised his son to rely on good deeds rather than on abundant offerings to the gods: "Do not be evil; patience is good. Make your memorial last through the love of you. . . . More acceptable is the character of a man upright of heart than the ox of the evildoer." The gods, he said, judge the dead on the basis of good deeds.[2] Thus the same kinds of problems and the same attempts at solutions present themselves, even though in different forms, from the beginning of history and under all political regimes.

Social upheaval and social concern were also common five centuries ago. Europe was then being devastated by endless conflicts—national, civil, religious. But knowledge and resources were increasing nevertheless as a result of explorations, scientific discoveries, and technological advances. At the time it did not seem that the social order could long survive the simultaneous impacts of wars and new forms of wealth.

In 1575, four centuries before the phrase "future shock" was coined, there was published in Paris a small book by an author who worried, as does Toffler now, about the disturbances caused by the new knowledge and technologies of his time, and even more about what was in store for the future. The time was the late Renaissance and the author was Louis Le Roy, a scholar and jurist so wise and learned that he was referred to in his time as the French Plato. In his book *De la Vicissitude ou Variété des Choses dans l'Univers* Le Roy presented the history—as he saw it—of technological developments since the beginning of time, but his main concern was the effects that the new learning and inventions were having on human life.[3] Le Roy's *Vicissitude* must have struck a sensitive nerve when it appeared, judging from the fact that it was reprinted in Paris six times between 1575 and 1584 and was translated into Italian and English respectively in 1592 and 1594. The English title was *Of the Interchangeable Course and Variety of Things in the Whole World*.

Le Roy indeed lived in the wake of exciting events The Renaissance had brought about profound changes in religious beliefs and in social patterns—changes which gave reason for both hope and despair. The feats of the great navigators and the geographical discoveries of the time were fully as exciting as the ability to fly around the world and the beginnings of travel in space. But Le Roy was disturbed by the knowledge that contact with new populations had exposed the explorers to new kinds of diseases. Syphilis had spread like a conflagration throughout Europe, as if introduced from the West Indies by Columbus's crew. Furthermore, its prevalence and that of other venereal diseases had been increased by the general relaxation of sexual mores and by the mass movements of armies and people at the beginning of the sixteenth century.

Other social changes were resulting from the fact that the printing press had made new information and ideas widely available. This had generated social crises throughout Europe by creating confusion in the beliefs of the public. The ancient loyalties which had bound the social order until then were breaking down under the impact of the new enlightenment, of the intellectual freedom, and especially of the religious conflicts. To make things worse, the introduction of gunpowder and firearms had rendered ancient weapons obsolete; wars had become more deadly and more destructive of property.

Le Roy had therefore good reasons for believing that an era of darkness and perhaps of self-destruction might be at hand. Despairing of his times, he exclaimed, "All is pell-mell, confounded, nothing goes as it should." Indeed new trends did seem to be incompatible with the maintenance of an orderly social structure. A century earlier, the dissolute life of François Villon had begun to give glamour to the social misconduct of the angry young poets. A number of women had attempted a feminist liberation movement. Nationalist spirit was spreading throughout the world: Peking had become the capital of China; the Turks had conquered Christian Constantinople; the Castilian kings had pushed the Moors out of Spain. As Charles de Gaulle was to proclaim the immortality of France even while the country was occupied by the Germans, so in the fifteenth century Jeanne d'Arc had incarnated the French spirit and thus helped to make nationalism the new religion of Europe. This continuous turbulence in all aspects of life occurred more than five centuries ago.

In 1900, after visiting the Galerie des Machines at the Paris World's Fair, the American historian Henry Adams became convinced that the era of the Virgin had come to an end and would be followed by the era of the Dynamo which would eventually destroy classical culture. He visualized a world in which the factory would replace Chartres.[4] The world of things has indeed continued to move forward at increasing speed and in an unpredictable direction. In 1927, it took Lindbergh's *Spirit of Saint Louis* twenty-four hours to cross the Atlantic; *Concorde* can do it in less than three. Weeks elapsed before the news came that Admiral Robert E. Peary had reached the North Pole in 1910, whereas we could actually watch astronaut Neil Armstrong and hear his voice as he set foot on the moon in 1969. Wars and murders now penetrate our living rooms instantaneously by television, and we can witness the birth or explosion of celestial bodies almost as readily as that of political regimes on earth. Various categories of experts predict that science will soon be capable of radically changing the surface of the earth, the structures of society, and even the genetic code of the human species.

Like our ancestors, we shall therefore inevitably experience traumatic changes in our daily life, and we shall long for the worlds we have lost. But it is only lack of historical perspective which leads us to believe that we are entering a period of upheavals which has no precedent. History is replete with examples of populations and civilizations which have been destroyed because they have been unable, or perhaps unwilling, to adapt themselves to new ways of life. History also shows, on the other hand, that many civilizations have taken advantage of their ordeals to renew themselves and thus become more creative.

Whatever their level of sophistication, all social groups have experienced crises whenever they have had to change their ways of life at too rapid a pace. During the nineteenth century, future shock manifested itself with special violence among certain primitive populations when they first came into contact with Western civilizations.

When the English settled in Tasmania, the island was occupied by a population which lived by hunting, fishing, and the gathering of wild plants. Life was easy and relatively happy for the early Tasmanians—at least if one believes the accounts left by Captain William Bligh (1788) and Capitaine Baudin (1802), the English and French navigators who discovered them. The Tasmanians did not use clothing but decorated their bodies with bright feathers and shells. Their tribal organization was limited to the authority of a wise elder or of the most skillful and bravest hunter; they were polygamous; they loved to sing and dance, especially to celebrate spring.

After the arrival of the English colonists, however, they were progressively dispossessed of their land and of their hunting rights. In order to avoid violent conflicts, the English government decided in 1835 to move them to Flinders Island, some 50 miles north of Tasmania proper. They were provided with decent housing, good clothing, and healthful food, but they were compelled to adopt new ways of life based on a narrow view of Christianity which had no meaning for them. Furthermore, they had to reorganize their lives according to a mercantile economy which clashed with the traditions they had developed as free hunters and fishermen in the forests and on the beaches of Tasmania. Despite the well-meaning effort of the English missionaries and civic authorities, the Tasmanians never became adapted to these new conditions of life and progressively fell victims to despair, alcoholism, and disease. The last survivor, a woman named Truganini, died in 1876, three-quarters of a century after the tribe's first contact with white people.[5]

Similar tragedies have occurred among Polynesian peoples. When Captain James Cook first reached the Hawaiian Islands, he found there a vigorous Polynesian population organized as a complex and prosperous society. But in Hawaii and also in Tahiti the Polynesians proved immensely susceptible to certain diseases introduced by the Europeans—scarlet fever, measles, tuberculosis, and syphilis, among others. Their biological and social collapse was accelerated by alcoholism, an affliction which commonly accompanies the breakdown of traditional societies. These pathological effects of the first contacts with Western civilization have now lost much of their virulence. But although the Polynesian people have now acquired resistance to the infectious diseases of European origin and to alcoholism, they still find it difficult to accept the styles of life and work introduced by American civilization into their islands. Other ethnic components of the Hawaiian population—those of Chinese, Japanese, and Philippine origin—have rapidly adapted to the technological ways of life, probably for historical reasons. But part of the Polynesian population seems to have lost the will to live—and this can be as real a cause of death for a collectivity as for an individual person.

The example of the Iks mentioned previously and similar tragedies which have affected primitive people in other parts of the world can be cited as evidence of the destructive effects of future shock. Many are the social groups for which a sudden change of conditions has been fatal. Many are the populations which have almost forgotten their glorious past, as is the case for the Mayas, the Incas, and certain people of the Near East. But other civilizations have survived terrific ordeals and have indeed derived new strength and new inspiration from this experience.

During the Old Stone Age, the great human innovations of the last glacial period emerged chiefly in the regions of the cold steppes and tundras, rather than in the milder climates to which the human species was biologically adapted. The progressive disappearance of the large game probably imposed more complex hunting practices, and the search for new resources compelled the Stone Age people to develop new kinds of social organization. Finally the need to find food by methods other than hunting led to the prodigious social changes which are summarized under the phrase "agricultural revolution."[6]

For thousands of years the Old World lived through many forms of civilization. In Western Europe, life has been repeatedly transformed—by the migrations of people from the east, then by the Roman domination, then again by the raids of Saxon and Germanic people. Christianity in its Roman, Byzantine, and medieval forms; the Renaissance and the first phase of the Industrial Revolution; countless national, civil, and religious wars have caused a continuous transformation in beliefs, patterns of thought, and ways of life. Moreover several of these transformations have occurred almost explosively, within less than one generation. Only a few decades were required to pass from the monarchy of divine right to the various forms of so-called democracy, from the era of stagecoaches to that of the railroad, from the shop of the artisan to the automated factory. In our time, future shock has taken even more spectacular forms—during the revolution that shook the world of the Russian czar and his muziks, and during the upheavals that converted the China of mandarins with long braids into that of Mao Tse Tung's short-haired communitarian society.

Prehistory and history thus seem to demonstrate the

fact that there is nothing permanent in the social forms of human life except their fundamental principles. But what seems obvious also is that the changes that have occurred have practically always involved deliberate choices. These choices have been the cause of many tribulations and of much suffering, but they have provided the environments in which humankind is constantly reborn.

2 / The Camp and the
Open Road

Despite its restlessness, humankind is at heart sedentary. People are comfortable and function well only in environments of which they know the resources and the dangers and to which they are emotionally adapted. Even with the vastly increased mobility of the modern world, most human beings today live and die within a few miles of the place where they were born. The vaunted mobility of the present time does not imply quite as profound a change as believed. It commonly means that the worker or the traveler moves to a place which differs geographically from his point of departure but where he finds working and living conditions similar to the ones he has left. For each *coureur des bois* who became an integral part of the New World to the extent of adopting Indian life and marrying into a tribe, there were millions of immigrants who converted the forest and the prairie into pastures and farmlands similar to the ones they had left and who intermarried as they would have in the Old World. For every true explorer who takes the plunge into an environment really new to him, there are thousands of colonists or tourists who re-create wherever they go the social environment they have always known.

The hunter-gatherers of the Old Stone Age generally established their camps near bodies of water and in locations where food was abundant. They traveled great distances at times, especially to follow the game, but they rarely abandoned their camps where they usually left women and children. These long associations with a given region, a camp, or a shelter are probably at the origin of the very human tendency to develop emotional attachments which can be so strong as to constitute an organic bond to a particular place. In all languages, this attachment is conveyed by expressions loaded with emotional values—at home, *chez soi, a casa, heim*. The effective quality of words like "the hearth," or in French, *"le foyer"* does not come only from the fact that the word evokes the place where the fire was kept going but also from the intimate relationships that the dwellers in a particular site established with the other members of their clan and with the objects which were part of their daily life.

During the Old Stone Age, certain sites close to one another sheltered for long periods of time human groups which differed in the styles of their tools, weapons, and other artifacts. The occupation of each particular site by a particular group thus had a certain permanency, even though the occupants must have moved away on many occasions either because of seasonal changes or simply to hunt and to gather food. This permanency of occupation implies that the "property" rights were recognized, even in the absence of the usual occupants—perhaps as a human expression of territoriality. The Indians of North America considered that the occupants of a particular site or territory had an absolute right to it as long as these occupants utilized it for their own ends. But they could not comprehend the European concept of land property considered as merchandise. This difference of attitude was responsible for countless misunderstandings and tragedies.

The symbolic significance of the hearth must have deep roots in the beginnings of magic, and it has spread throughout most forms of religion and social structures. The site where fire was made or maintained was perhaps at first a cave, as for example the Choukoutien cave in China. However, *Homo erectus* built artificial shelters as far back as 300,000 years ago. Some of Neanderthal man's constructions were so large and complex that they implied sedentary life. Strong houses of stone and clay were built 10,000 years ago in Palestine and in the upper Euphrates valley—evidence that the stability of habitation preceded the emergence of agriculture. The emotional values associated with the hearth have therefore long been associated with the permanency of habitation.

While early people were fundamentally sedentary, they must at times have been compelled to abandon their sites as a result of population increase, of competition for hunting or gathering territories, and, on a longer time scale, of climatic changes. As already mentioned, migrations had spread humankind over the whole earth by the end of the Old Stone Age. A vague memory of these dispersals seems to have persisted in the legends of most primitive people and also in the form of a nostalgia for the lands of the ancestors. Few are the people who do not have a racial memory of Arcadia, if not of Eden. Yet hope for a better world—beyond the hills or the ocean—seems to have generally coexisted with nostalgia for the lost paradise. For almost a hundred centuries, people from Asia and Eastern Europe have migrated toward the Atlantic and Mediterranean shores, then have spread over the rest of the world. Since the seventeenth century, European people have systematically colonized the earth and have dreamed of moving into outer space or of settling on the bottom of oceans. In modern bourgeois societies, the craze for tourism may be another expression of the desire for movement which began with the Stone Age dispersals.

Many of the great migrations occurred without a real utilitarian purpose. Throughout history, large bodies of populations have traveled on pilgrimages, crusades, or holy wars under a variety of religious or other banners. Part of the motivation, however, was certainly the need for adventure and the desire to search for the treasures that imagination commonly associates with the unknown. It is not only gold that attracts humankind to Eldorado, but also the hope of discovering the world of its dreams. In the Western world at the present time, the majority of migrations are directed to the lands of the sun and to a few privileged coasts. These are the regions most favored by tourists, and those where people like to work or retire. It would seem that humankind, after having achieved its highest development in harsh climates under strenuous conditions, is now tired of its efforts and longs for the subtropical environments in which it had its biological origins. The migrations toward the sunny coasts thus do not correspond to the search for adventure, but rather to the hope of rediscovering the lost paradise. The safaris to Africa or the Club Méditerranée in Tahiti give humankind the illusion that it can recapture peace of mind and body by re-establishing contact with natural conditions in environments physically adapted to primitive human nature.

In recent times, the rejection of sedentary life has

been most evident in the United States. Jack Kerouac's *On the Road* expressed the beatnik's horror of taking root—anywhere—for fear he would have to accept the constraints of society. Kerouac's words, "Where are we going, man? I don't know, but we gotta go,"[7] speak to all those who wish to escape—and who does not long to escape at certain moments of his life?

In all industrial countries people of all ages dream of returning to unspoiled nature, there to start a new life in intimate contact with the land. For a few adventurous people, this means life in the wilderness or at least in communes. For the more conservative, it simply means spending as much time as possible in a secondary home under somewhat rustic conditions. The desire to take roots has brought about a revival of regionalism. The various parts of each nation are trying to give new life to their traditional ways; folklore is fashionable. The different sections of large cosmopolitan centers proclaim their individuality; instead of being a melting pot America becomes a mosaic of ethnic groups. Every part of the world fights for its national identity and for independent membership in the United Nations. Electronics may create the illusion that the earth has become a global village, but in fact most people feel really comfortable only when at home. However, the longing for change and adventure is paradoxically also a universal component of human nature.

3 / Ulysses and the American Frontier

Myths and legends have a core of unchangeable truth but, like rolling stones, they change shape with time. Also like stones which acquire local moss when they come to rest, myths and legends take on new characteristics at each period of history. They can thus remain meaningful and timely by adapting themselves to the moods of the contemporary world. About three thousand years ago, Homer codified in the *Odyssey* some Mediterranean tales which were even then extremely ancient. He thus launched on its course the legend of Odysseus, his travels and his tribulations, which has since undergone many mutations in European literature. But the legend still symbolizes the adventures, ordeals, and temptations all human beings experience as they try to fulfill their destiny, whether in the Aegean Sea or in the streets of Dublin.

Homer's Odysseus was obviously a complicated person with many conflicting traits. He loved adventure and could withstand extraordinary hardships, but he was also sensitive to the charm of women and to the beauty of the wine-dark sea. He fell readily under the spell of Circe and Calypso and could enjoy the ease of life among the Lotus-Eaters. But he had constantly one goal in mind: he wanted to return home to Ithaca. He gave up the chance to enjoy perpetual youth with Calypso for his uninspiring duties as a landowner and the companionship of aging

Penelope. In the mind of the Greek poet, the fulfillment of human life thus implied both the excitement of adventure and the return home to one's origins. No traveler can forget his past: the appeal of the road does not erase in his memory the image of the place from which he started.

For later European poets, Odysseus's travels have symbolized the search for knowledge and for spiritual values. The belief that human life can reach beyond animal nature is personified in the character of Ulysses as described by Dante in the *Divine Comedy*. On his journey, Dante's Ulysses reached, at the Pillars of Hercules, a boundary stone which instructed him not to proceed farther. Yet he exhorted his companions to go on. "You have your lives, not so that you may live like beasts, but rather that you may strive for fame and knowledge." Dante's own view of man was based, however, on the static concepts of Christianity at the end of the Middle Ages; he believed that the search for truth should remain within the framework of the Holy Scriptures. In the *Divine Comedy*, Ulysses and his mariners encounter a violent storm which swallows up their ship. Dante thus points the moral that humankind will founder on its presumptuousness if it transgresses the limitations set by God.

In contrast to Dante, Alfred Tennyson believed, as was almost universal in the nineteenth century, that there are no limits to scientific knowledge and therefore to the growth of civilization. When the English poet restated Homer's theme in his own poem "Ulysses," he took it for granted that progress is open-ended and that its pursuit is the very essence of the human condition. His Ulysses proposes to sail "beyond the sunset" with the conviction that humankind will inevitably achieve a better estate just by going forward .

The immense poem, *The Odyssey: A Modern Sequel*, by the modern Greek poet Nikos Kazantzakis reflects the euphoria and also the depths of pessimism which have characterized the twentieth century. His hero returns successfully to Ithaca and settles on his domain as Homer's Odysseus had done. However, he soon becomes bored with his eventless existence and sails once more away from Ithaca, this time for good. With the help of a few companions who share his restlessness, he starts in search of adventure and of the true life. But despite extraordinary experiences in all the citadels of knowledge and of philosophy, nothing positive comes out of his travels. He concludes in the end that there is no intellectual or ethical meaning to life; only art, perhaps, will survive. He had not found satisfaction at home and he had become aware of the fact that adventure does not lead anywhere.[8] Certain American poets, even more pessimistic, write that the open road leads to the parking lot or to the garbage dump.

The deep significance of Ulysses's story, in its many variations, is that the vicissitudes of life come in large part from conflicting tendencies inherent in human nature. Taking to the open road gives many satisfactions but we never forget our origins and for this reason we want to return to camp. On the other hand, sedentary life is not completely satisfying either, because we are never perfectly adapted to our social environment, and furthermore we yearn for adventure. We thus constantly oscillate between the two ideals of being and becoming. These hesitations are responsible for much human uneasiness. They are creative, nevertheless, because they engage human beings in a process of self-searching and of external exploration which helps them to escape from biological bondage and to humanize the earth.

The clearing of forests and the draining of marshes had completely changed Europe by the end of the Middle Ages. In North America, a similar process took place between the seventeenth and the twentieth century, converting most of the continent into a huge agricultural and industrial enterprise. The picturesque phrase "the breaking of the prairie" conveys the immense effort that was required for this achievement. There were profound differences of attitude between the settlers of the European and of the American continent. During the Middle Ages in Europe, the land was cleared by a native population which continued to live and work on it. In contrast, the transformation of wilderness in North America was largely the feat of people who had recently come from other parts of the world and who in many cases did not settle permanently on the land they had cleared. The peculiar social conditions associated with the rapid spread of human occupation from the Atlantic to the Pacific coast had consequences that can still be recognized in the American temperament.

Until the very end of the nineteenth century, the westward movement in the United States was marked at its advancing edge by a zone in which the human settlements were still very primitive; this is commonly referred to as the frontier. This usage of the word "frontier" is peculiar to the United States and confusing for non-American readers. It does not apply to a geographical line of demarcation between countries or regions, but rather to the levels of technological development, the social structures, and the ways of life that prevailed in the areas where the process of settlement was still incomplete. In this ever-moving "frontier," people coming from the American east coast or from other countries had to cope with the primeval forest or the prairie which had until then been occupied only by a sparse Indian population. Moreover, contacts with Western civilization were difficult and rare. The word frontier thus meant an early phase of social development rather than a geographical area. During this phase, almost everything had to be hastily improvised—from the ways of life and of work to the political structures and to the solution of conflicts with the Indians. This required a range of attributes—strength, courage, initiative, ruthlessness—which are now summarized by the expression "frontier spirit."

Just before the turn of the century, the American historian Frederick Jackson Turner (1861–1932) formulated the theory that the enterprising spirit of American people and the nature of their institutions have their origin in the experiences of the early settlers as they progressively moved from the Atlantic to the Pacific coast. "The frontier," wrote Turner, "was a gate of escape from custom, class restraints, economic and social burdens." It determined many of the distinctive traits commonly regarded as characteristic of American life: strength with coarseness, practicality and inventiveness, restless energy and materialism. The frontier experience developed in Americans the skill to work out practical solutions for the new problems they were encountering. It instilled in them a profound sense of democracy and it shaped their fundamental optimism by giving them the illusion that they could forever continue to move forward.[9]

Historians have come to doubt that the role of the frontier in American history was as formative as Turner considered it. The legends of Western life notwithstanding, most settlers tended to be rather conservative in their social and political attitudes; they contributed less to the development of democratic institutions than did people in the older urban areas. It is even questionable that the settlers were greatly innovative in technological matters. The Springfield rifle, the Colt revolver, and the covered wagon were developed not in the Western settlements but in Eastern cities. To a large extent, the frontier was therefore the destination of things and ideas rather than their source.

Granted that the frontier experience was only one of the many factors which shaped American history, it is certain nevertheless that the American psyche has been deeply imprinted by the frontier myth. Stories are still growing about the daring and energy that were displayed in the winning of the West. The qualities that made possible the clearing of the land and the conversion of forests into lumber, of ores into metals, of river banks into wharves in such a short time are said to explain modern American efficiency in converting all natural resources into urban agglomerations, industrial areas, and super-highways. Until recent times, America was almost universally regarded as the wide-open land of unlimited opportunities, where the strong self-reliant individual could thrust his way to the top.[10]

There has not been much open land in the United States since the beginning of the twentieth century, and natural resources are no longer as readily available as they used to be. In fact, the westward migration which occupied Americans for more than two centuries had essentially ended when Turner first presented his frontier thesis in 1893, and he himself realized that a major change in national psychology was thus inevitable. But myths survive long after the disappearance of the conditions from which they originated. When Americans were compelled to adjust their economy, political structures, and daily lives to a closed-space world, they transferred the frontier myth from geographical expansion to industrial development. In 1945, the American scientist and technologist Vannevar Bush gave the myth a new lease on life with his book *Science, the Endless Frontier*, which stated that scientific knowledge could become the chief source of technological wealth just as natural resources had been in earlier times.

Vannevar Bush is the spokesman for the view that the spirit of exploration and adventure can now best express itself through scientific research and technological applications. To use his own word, this new kind of frontier is "endless" because there are no limits to scientific exploration.[11]

However, attitudes and convictions change rapidly, even in scientific matters. Some philosophers and a few scientists now reject the view that there are no limits to scientific knowledge. In the course of a symposium organized in 1973 by the National Academy of Sciences to celebrate the five-hundredth anniversary of the birth of Polish astronomer Nicolaus Copernicus, one of the participants dared to suggest that the "modern scientific era which began with Copernicus may be reaching its end, stopped by the same forces that brought to an end the golden age of Greek science." More important, there is a widespread feeling, even among scientists, that science and technology are taking humankind on a dangerous road.

A simple comparison will suffice to show how deep

is the change in attitude with regard to science and progress which has occurred during the past quarter-century. Vannevar Bush was president of the Massachusetts Institute of Technology when he published *Science, the Endless Frontier*. In 1972, young professors of the same institute achieved instantaneous international fame by publishing a book entitled *The Limits to Growth*, in which they claim that technological growth cannot possibly last more than a few more decades longer and that the continuation of present forms of scientific technology would inevitably lead the world to disaster.[12] The same year a group of illustrious British scientists expressed approval of a manifesto, *Blueprint for Survival*, which affirms that humankind's only chance of survival is to abandon the pursuit of technological progress as presently conceived. The solution proposed in *Blueprint for Survival* is to adopt ways of life based on the use of local resources, both for agriculture and technological production.[13] Thus, sophisticated scientists and sociologists have joined humanists in rejecting the traditional concepts of progress and in questioning the value of the forward march through scientific knowledge which has heretofore been identified with Western civilization. According to them, the time has come to make camp.

All wise people of all times have felt, like Voltaire's Candide, that the surest way to happiness is to cultivate one's garden. Also like Candide, however, they have come to this conclusion only at the end of a turbulent life, and commonly because they have had their fill of adventure. If material comfort had been the dominant value in the life of *Homo sapiens sapiens* he would have settled once and for all in a pleasant semitropical climate; instead, either by force of circumstances or by deliberate choice, he moved all over the globe. He thus broadened his knowledge of the earth, his spectrum of perceptions, and his range of choices and so achieved the social evolution which accounts for most forms of civilized life.

By taking to the road, literally and figuratively, our ancestors exposed themselves to conditions that have enabled them to evolve both biologically and socially and that enable us now to continue exploring and developing the potentialities of the human species.

4 / Technologic Utopia

"For centuries, humankind has been moving forward—as in a dream." This sentence, which reflects the euphoric attitude of the Western world at the end of the nineteenth century, is the free translation of two lines in a poem by Victor Hugo: *"Depuis des siècles, en révant / le genre humain marche en avant."* The dream to which Victor Hugo alludes is that of a world in which the human species would have achieved complete domination over natural forces and thereby become master of its own destiny. It is a very ancient dream, at least as ancient as the legend of Prometheus.

A gentle ecological romanticism encourages us to assume that primitive populations lived in harmony with nature. According to this assumption, the Old Stone Age hunters identified themselves completely with the animals and plants in their surroundings, and they never engaged in actions which alienated them from their environment. But prehistoric life was certainly very different from this image. For several thousand centuries the precursors of the human species have behaved as if they were intent on dominating nature—first by the use of fire and simple tools, then with ever more complex and powerful technologies. The technological equipment of the Cro-Magnon people was so sophisticated that it enabled them to trap and kill almost any kind of animal and to create almost anywhere on earth the artificial shelters they needed. At the beginning of the Neolithic period, the techniques of deforestation and of irrigation were sufficently developed to permit the creation of the man-made lands we now call nature. Thus technology had become an integral part of human life long before history began.

After the seventeenth century, inventors increasingly used theoretical science to improve the design and operations of the machines first created by the skills of artisans. Scientific technology rapidly became so powerful that technological civilization appeared to be capable of uninterrupted progress. In 1950, coal, petroleum, and other fossil fuels were readily available and provided at low cost all the energy required for industrial processes, while techniques for the utilization of nuclear energy were so advanced that many people stopped worrying about the depletion of fossil fuels and of other natural resources. It appeared that human beings needed only imagination and daring to invent the future they wanted and bring it into being.

The technologic euphoria which began about 1600 with Francis Bacon and was continued by the eighteenth-century philosophers of the Enlightenment achieved its most extreme expression among the twentieth-century futurologists, who took it for granted that the year 2000 would see the dawn of a technologic utopia. Everyday life in the first half of the twentieth century seemed indeed to justify the confidence of utopians, because the achievements of scientific technology commonly went far beyond the most optimistic dreams of the experts. The magnitude of recent achievements can be realized from an anthology made up of articles published in American magazines around 1900.

At that time, public opinion was excited about the prospects of what life would be like during the twentieth century, and the anthology therefore gives an entertaining opportunity to compare what was then imagined with what has really happened.

In 1900, the general public marveled at the great steamships that crossed the Atlantic in ten days, and it was fashionable to imagine even larger and more luxurious steamships with more and taller smokestacks, so rapid that the crossing from America to Europe would be cut to less than five days, but none of the experts even suggested that jet airplanes would cross the Atlantic in a few hours. The decades before 1900 had seen the development of improved strains of vegetables and fruits so that the experts felt entitled to predict for the near future strawberries as large as melons, but none of them imagined quick-frozen foodstuffs or TV dinners. This was the era when women had begun to escape from the tyranny of Victorian fashions and prophets envisaged skirts so short that they would reveal the ankle, but even the boldest of Paris couturiers would have blushed at the thought of miniskirts, bikinis, or "hot pants."

Scientific technology and the change in tastes and social conventions have led the twentieth century in directions so different from what had been imagined by the 1900 experts that the tendency is now to imagine a twenty-first century where everything will be completely unlike what we have today. In 1960 futurologists predicted that physical and mental effort would become useless thanks to robots in the kitchen and to learning machines in the classroom; that suffering and disease would be eliminated by the use of drugs; that dreams would be programmed with electronic machines; that children would be produced artificially in the test tube and conditioned by manipulating their environment; that whole cities would be sheltered and air-conditioned under transparent domes; that permanent stations would be established on the moon or in the oceans to exploit new resources and to accommodate excess populations of earthlings. It was believed, in brief, that the physical and human world could be reshaped at will.

Much of this technologic utopia is within the domain of scientific possibility, and yet there is hardly any chance that it will come to pass, at least for several generations. One of the reasons is simply that human beings select among the possibilities offered by scientific technology and that what is possible does not necessarily appear desirable. Supersonic transport and "hot pants" almost went into oblivion the same year, not because of technological difficulties but for lack of public support. Another reason is the new awareness that technological innovations commonly have disastrous secondary effects, many of which are probably unpredictable. In the case of the automobile, it would have been impossible to predict half a century ago the nature of the pollution problems and social disturbances that it would cause, because no one could anticipate the extent of its use; life is enriched by one million automobiles but can be made into a nightmare by one hundred million.

There is almost universal agreement that vigorous and immediate action is needed to repair the damage already done to the urban and rural environments by the mismanagement of industry and agriculture during the past two centuries. This task will certainly require enormous amounts of material and intellectual resources, and interest in new esoteric technologies will be dampened accordingly. During the forthcoming decades, social efforts will be focused on trying to make the world as pleasant as it used to be, rather than on creating entirely new ways of life in artificial environments. After a phase of almost blind faith that progress is identified with technologies of greater and greater complexity, our contemporaries are beginning to rediscover the satisfactions that come from direct contact with reality through the senses and from sharing emotional experiences with a few people. These satisfactions have long been and still remain the surest approach to well-being and happiness, but they are the ones most threatened by the ill-conceived utilization of science and of technology in the Western world since the Industrial Revolution.[14]

Throughout most of its existence, humankind has believed in the stability of modes of life and social institutions. The notion of progress through rapid and continuous change has been widely accepted only during the past two centuries, first in the Western world and then progressively in other continents. But paradoxically, the general public is beginning to question the merit of technological progress precisely at the time when greater knowledge and means are available for accelerating its course. The change of attitude does not come from a rejection of the conveniences introduced by science but from the realization that each improvement is likely to have a high social cost. Industrial efficiency requires the mechanization of labor; increase in the quantity of production commonly limits diversity and decreases the interest of the product; airports and thruways facilitate travel but take the traveler away from the landscape. The new satisfactions that technological progress creates do not compensate for the loss of luminous skies, fragrant air, pure water, quiet and harmonious neighborhoods, waterfronts that appeal to the swimmer, landscapes where it would be pleasant to walk.

The present world-wide effort to save the quality of the environment transcends the problems posed by pollution and by the depletion of natural resources. It constitutes rather the beginning of a crusade to recapture certain sensory and emotional values, the need for which is universal and immutable because it is inscribed in the genetic code of the human species.

5 / The Incarnations of Humankind

One could, I believe, defend the thesis that social evolution has been accelerated and perhaps even motivated by the continuous efforts of humankind to escape from boredom through the pursuit of adventure. Even the most material aspects of human existence commonly take forms which enliven their tedious routine and endow them with spiritual values. Cooking is often a repetitive chore, but it has its creative aspects and it can transcend drudgery through uniting the family at mealtime. The preparation for a large dinner party is an exhausting task, but one which can acquire a ceremonial character if its purpose is to celebrate an event having deep emotional significance for the family or the social group. Sexuality is an expression of animality but, sublimated, it has greatly added to the romance of life. Every human group has converted biological necessities into cultural forms that are transmitted from generation to generation and thereby give social identity to the group.

Especially in the distant past, certain cultural forms and ways of life remained almost stable for long periods of time, as if primitive people had found complete satisfaction in the social structures transmitted to them through their ancestral traditions. This stability, however, is not a fundamental characteristic of human life; it is rather the consequence of isolation from the external world. All societies undergo rapid and profound changes when they establish—by choice or necessity—intimate contact with people of other civilizations. The eagerness to change is so widespread indeed that, whenever life appears too stable and comfortable in sophisticated societies, restlessness is expressed in a wish to return to primitive conditions. Théophile Gautier's phrase *"Plutôt la barbarie que l'ennui* [Rather barbarism than boredom]" expresses a desire, common in nineteenth-century Europe, to escape from the stability and stuffiness of bourgeois civilization. Unwittingly, the bourgeois establishment may have contributed to civilization by generating countercultures among the young and by motivating artistic rebellion.

The desire for change accounts in part at least for the large movements of populations which have constantly occurred since the Stone Age. There is a peculiar paradox in the direction of these movements. Whereas various Holy Scriptures place Paradise somewhere in the East, population movements have been chiefly westward-oriented. Western Europe has been repeatedly invaded by people of Teutonic, Scandinavian, Slavic, and Asiatic origins; Europeans have emigrated to the Americas; North Americans have progressively moved from the Atlantic Coast toward the Pacific Coast and now many of them cross the Pacific in search of enlightenment in Asia. Perhaps the symbolism of the westward march is that adventure, despite the hardships it entails, offers more desirable human values than does the condition of peace and comfort associated with the eastern Paradise.

The phrase "human values" is loaded with emotion,

but its intellectual content will remain obscure until agreement is reached concerning the attributes which are specifically "human." While the biological characteristics that define *Homo sapiens sapiens* are easy to describe and to measure, there are no scientific criteria for defining the humanness of the human species. Ever since the seventeenth century, science has concerned itself almost exclusively—and wisely so, in my opinion—with those aspects of reality which are governed by the deterministic laws of nature. But the deterministic approach has great limitations when applied to the study of humankind because it means that scientific knowledge is focused on the description in physico-chemical terms of a brutish creature deprived of feeling, and on the social analysis of masses of robotlike creatures which contribute to the Gross National Product without asking the uniquely human question of *why* they produce and consume. The distinctive characteristics of humankind obviously transcend those which can be studied by scientific methods. One is human to the extent that one remembers the past and is concerned about the future; that one manages to combine in one's personal life such different attributes as rationality, intuition, and feelings; that one can communicate with other human beings in such subtle ways as to identify oneself with their fate. In the final analysis, one is human to the extent that, while remaining an animal, one transcends those aspects of behavior which are deterministically governed by animality.

All civilizations have formulated ideals of behavior based on attributes which appear uniquely human. In each case, this ideal is the expression of forces originating, not from geography and race, but from a philosophical attitude; not from the animality in *Homo sapiens*

sapiens but from his humanity. The aspect of Chinese civilization symbolized by Confucius emphasizes the quality of social relationships. In contrast, the Taoist view expressed in Lao Tse's poem "The Way" considers humankind as a part of nature, but with expressly human feelings as to behavior and relations to nature. For Hebraic scholars, the most praiseworthy human attributes are dignity, rectitude, and a sense of responsibility. Still different arrays of qualities were associated with the phrase *"honnête homme"* in seventeenth-century France or with the word "gentleman" in nineteenth-century England. The forms taken by humanness seem to be as varied as the forms of artistic expression, although in both cases they probably represent the selected aspects of a more complex core of truth which is common to the whole human species.

The diversity of attitudes concerning ideals of behavior is only one of the factors which contributes to the spiritual richness of humankind. Another peculiarity of humankind, as already mentioned, is its power to change deliberately its habits, goals, and social structures. The profound changes which have occurred in the course of social evolution have been influenced, of course, by the accidents of history, but they have also reflected choices which are none the less voluntary for being at times subconscious. During the past 10,000 years, for example, Europe has experienced a succession of cultural patterns so different one from the other that they can be identified by such simple phrases as the Neolithic Revolution, Imperial Rome, Roman Christianity, Byzantine Christianity, the Middle Ages, the Renaissance, Protestantism, the Enlightenment, Romanticism, the first Industrial Revolution, the Technological Age. In each case, the shift from one form of civilization to another was extremely rapid and was to a large extent the outcome of conscious efforts.

The technologic formula of civilization dominates the world at the present time, but there is no guarantee that it will be permanent. In fact, it may not last longer than the preceding forms of Western civilization; it may be rejected outright or be so profoundly modified that it will appear to our descendants as a barbaric phase of history. Instead of technology's operating almost as an independent force as it does today, it may be incorporated in a more human order of things so as to be less destructive and more creative of real values.

On several occasions in the past, humankind has tried—admittedly with little success so far—to formulate doctrines that would enable it to achieve harmony with the cosmos. The Buddhists put their faith in the philosophy of the Nirvana and the Judeo-Christians in the vision of a personal God. In our times, orthodox ecologists believe in the possibility of perfect equilibrium in natural systems and reject the hypothesis that the human species occupies a special position in these systems. But a large percentage of human beings are of the opinion that humankind is responsible for the stewardship of the earth and that this implies not only acting as caretaker of the past but also as creator of new environmental values.

Whatever its failings, the present technologic age can be regarded as one of the most brilliant periods in the history of humankind. However, the increasing frequency of protests against technology reveals that it has failed to satisfy certain human aspirations. It has made life easier and on the whole safer, but many people believe that it has made the world less interesting and has increased frustration and boredom.

To speak of boredom seems contrary to commonsense at a time when the public exhibits enthusiasm for one cause after the other—for African art, space travel, pop music, racial problems, environmental quality, oriental mysticism, and so on and so forth. But this rapid succession of fads implies of course that each of them is soon abandoned and forgotten. Such lack of real commitment may contribute to the ennui—the organic sense of boredom—which is most common in prosperous societies where the harsh struggles for daily life have been eliminated and where so many new and interesting experiences are available constantly. It is hard to believe and yet true that interest in space exploration is already on the decline and that even the names of the moon explorers are being forgotten. Space travel and voyages to the moon have been a human dream for several thousand years, but paradoxically they are losing their romantic appeal precisely at the time when they have become reality. In contrast, the experience of daybreak or sunset has retained all the power it has had since the beginning of time, probably because it can be enjoyed without instruments, directly with the naked eye. The rapid loss of public interest in space exploration may have its origin in the fact that even the astronauts have only limited sensory contact with the world they explore. Enclosed in the artificial atmosphere of space capsules or space suits, which duplicates some of the physicochemical properties of the terrestrial environment, they are limited in the real experience of the *terra nova* they traverse. They cannot perceive it through all their senses and therefore cannot communicate it in terms which are meaningful to the general public.

Man knows the external world and responds to it

only through its impact on his senses, and indeed on his whole body. His flights of imagination are powered by perceptions which have a purely organic source in the physical world. Even the theoreticians of science commonly find it useful to imagine concrete models of what they investigate before developing their abstract concepts of reality. In imagination they "see" the forms and structures of atoms, genes, or galaxies in shapes similar to those of the visible world, before working out the mathematical formulations which define the unseen structures they study.

For the immense majority of human beings, the only valid experience of reality is the direct sensory perception. This experience can be transferred from one person to another only if it can be described in images which are based on sensory perceptions common to the human species as a whole, and even then only within a specific cultural framework. In this regard, space explorations, spectacular as they are as technological achievements, constitute a failure of scientific civilization. They symbolize the power of scientific technology but also the progressive loss of direct sensory perceptions without which there cannot be any organic enjoyment of reality.

It is becoming more and more difficult in the technological world to enjoy the simple fundamental sensory experiences which were available to the Stone Age hunter, to the peasant by his fireplace, or even to the townsman before the era of gasoline and elevators. We have the advantage over our ancestors of knowing in the abstract more about life in distant countries, of being able to move far and rapidly by land, sea, or air, of watching a spectacle or hearing a concert in the comfort of our apartments. But whatever the level of material civilization, children continue to derive their chief pleasure from eating, running, and jumping; young adults continue to dream of romantic and sexual love; oldsters continue to seek quietness and to worry about death and afterlife. Technology transforms all the artificial aspects of existence but affects very little the activities in which one becomes really involved organically and emotionally. Social contacts may have been more satisfying by the fire in a Stone Age cave or on a village bench than they are now through the convenience of telephone conversations and of other means of mass communication. Dancing to the sound of drums in the savanna or of a fiddler on the village green could be as exciting as dancing to electronic music. Throwing a rock at an enemy was a more satisfying way to express anger or hatred than killing him at long range with a gun or a bomb.

The fundamental satisfactions and passions of humankind are thus still much as they were before the advent of the automobile, of the airplane, and of the television set; before the era of steam and electric power; and even before our ancestors had abandoned hunting for agriculture and for industry and had moved from the cave to the village or the city. In many cases, furthermore, modern life has rather impoverished the methods by which fundamental urges can be expressed. Modern societies can escape from boredom only by supplementing the values of technologic life with the direct sensory experiences of primitive life; the need for these experiences persists in the modern world for the simple reason that it is indelibly inscribed in the genetic code of the human species.

Now and then, groups of people attempt to recapture fundamental sensory experiences by returning to life on the farm under primitive conditons, but the attempt is rarely successful. The peasant's life is often so exhausting and leaves so little freedom that it approaches slavery. Just like the factory or office worker, the traditional peasant and even the modern farmer finds his pleasures outside of his trade—in festivities with abundant food and drink which help him to forget for a few hours the tedium and rigors of his working days. He goes to the town's Main Street whenever he can and travels to Florida or Las Vegas if he can afford it. Everywhere and throughout history the peasant has tended to abandon the farm whenever he had a chance, were it only to look for an easier and more exciting life in shanty towns.[15]

The novelists and poets who write in romantic terms of the simple peasant life are not the ones who plow the land or take care of the cattle. It is true of course that country life is rich in sensory and emotional values. But such values are experienced less by those who have to farm for a living than by those who can enjoy nature without having to enslave themselves to the routines of agricultural life. In his *Meditations on Hunting*, the Spanish philosopher José Ortega y Gasset describes lyrically the rich perceptions and the intensity of experience associated with hunting practices:

"When one is hunting, the air has another, more exquisite feel as it glides over the skin or enters the lungs, the rocks acquire a more expressive physiognomy, and the vegetation becomes loaded with meaning." The hunter, Ortega continues, ". . . will instinctively shrink from being seen; he will perceive all his surroundings from the point of view of the animal, with the animal's peculiar attention to detail. This is what I call being *within* the countryside . . . Wind, light, temperature, ground contour, minerals, vegetation, all play a part; they are not simply there, as they are for the tourist or the botanist, but rather they *function*, they act. And they do not function as they do in agriculture, in the unilateral, exclusive, and abstract sense of their utility for the harvest, but rather each intervenes in the drama of the hunt from within itself."[16]

I have quoted Ortega at such length, not for what he says about hunting per se, but because he expresses so acutely the richness of sensory experiences that can be derived from direct contact with nature—human values that are being eliminated by technological civilization. It is fashionable at present to argue that pollution and depletion of natural resources will soon be the limiting factors of technology and economic growth. In their evaluations and forecasts, however, economists and futurologists never mention the values discussed by Ortega, even though such values are certainly essential to a full life. It would be out of place to discuss here problems of social organization, but it is worth suggesting that one of the "limits to growth" may turn out to be the shortage of fundamental satisfactions such as were provided by simpler ways of life. In order to retain their sanity human beings must try to recapture these satisfactions because the need for them is an unchangeable part of human nature.

When considered as a member of the animal kingdom, *Homo sapiens sapiens* exhibits a remarkable biological stability. When he is considered from a purely human point of view, however, cultural and social changeability becomes his dominant characteristic. The social and cultural mutations which provide the stuff of history are the successive incarnations of human potentialities. These multiple incarnations, which we call civilizations, reflect the ways that different aspects of the environment are perceived through the senses and organized by the human mind.

6 / Revolutions and Resurrection

To leave one's mark on the surface of the earth used to be a personal adventure with poetical qualities, but it has now become anonymous and banal through the complete mechanization of all enterprises. The lone woodsman has been displaced by highly organized teams of lumberjacks; one no longer hears the deep rhythmic sound of his ax in the forest, only the grinding shriek of mechanical saws. Instead of country people progressively opening the trails from village to village, or Roman conquerors deciding on the course of the roads to be built for their legions, bulldozers now cut through mountains and level the ground for thruways largely designed by computers. But while the methods for the transformation of the earth have become immensely more powerful and more rapid, the effects of these transformations are still poorly understood and largely unpredictable. To a large extent the environmental crisis has its origin in this lack of understanding and of predictability.

The distinctive characteristics of each civilization have resulted from the interplay between certain traits of human beings and the environments in which they develop and function. In the course of this interplay the different human groups have modified their environments which in turn have molded human nature. The histories of England and France provide striking illustrations of these reciprocal effects.

Beast or Angel?

The Vikings who settled in French Normandy at the beginning of the Middle Ages and from there occupied England after the battle of Hastings rapidly became one of the dominant forces of Western civilization. Their attitudes and ways of life changed profoundly and rapidly as they spread through various parts of Europe and, during the Crusades, the Near East. But it can be taken for granted that the genetic structure of the descendants of the Norman and English barons remained much the same as that of the original Vikings who sailed from the Scandinavian coast more than ten centuries ago. Yet the civilizations the Normans shaped on the two sides of the English Channel soon acquired distinctive characteristics with regard to social structures, ways of life, and even the appearance of the land.

East Anglia and French Normandy have approximately the same climate, topography, and soil composition. But they have become regions which differ in their types of land ownership, agricultural practices, natural scenery, and landscape architecture, even though they have been dominated since the beginning of the Middle Ages by people of the same Scandinavian origin. The differences between the two regions influenced in their turn the behavior and social evolution of the people involved. Despite continuous exchanges between England and France at all levels of commerce, technology, science, and culture, the English and the French people clearly differ in their psychological characteristics. When Winston Churchill stated before the House of Commons, "We shape our buildings and afterwards our buildings shape us," he could have used as illustrations the profound differences in habits and tastes between the people on the two sides of the Channel.

Churchill's phrase expresses a biological truth which is as valid for other living things as it is for human beings, in the sense that animals and plants are also shaped by the environments in which they develop. But humankind has introduced a new factor in the manifestations of environmental conditioning. Whereas animals and plants become adapted to new environmental forces through unconscious genetic mechanisms, human beings consciously modify the environment in which they live to fit their needs and even more their whims and aspirations. The various human groups thus affirm their identity in the face of natural forces, which they use to enrich their beliefs and their social habits. If East Anglia now differs from French Normandy, this is not due to geographic differences between the two regions, or to genetic differences between their inhabitants, but to social forces which are themselves the consequences of human choices and decisions.

For more than a hundred millennia different human groups have thus followed their respective destinies, utilizing natural forces to create the man-made environments in which they function and which we now call nature. This process, which has transformed a large percentage of the surface of the globe into humanized nature, has deep roots in prehistory but may have nevertheless begun through intellectual perceptions. As early as the Old Stone Age, people had recognized, and noted by markings on bone or ivory, certain regularities in the movements of the sun and the moon, in the rhythms of the seasons, and in the life processes. The awareness of this natural order accompanied them in their migrations and helped them to create an order of their own out of the environments in which they found themselves—in other words to create their own man-made order of nature. The various human groups have evolved socially along innumerable channels, thus developing their distinctive civilizations. Many of these have died on the way; others have reached dead ends and remained static for long periods; a few have continued to move on even when they suspect that they are on a dangerous road. As Paul Valéry stated in a famous phrase: *"Nous, civilisations, savons maintenant que nous sommes mortelles* [We civilizations now realize that we are all mortal]."[17] This phrase is in fact applicable to all living species as well as to civilizations, and it may even have a stronger meaning for animal species than for the human species.

The processes of Darwinian evolution are essentially irreversible. Once an animal species starts to develop in a certain evolutionary direction, it has little chance to escape. It becomes slave to the natural forces and ways of life that have shaped its biological characteristics and its habits. The gorilla is condemned to an herbivorous way of life in a certain type of forest; the lion is condemned to the life of a predator in the savanna type of country; each of the species of social insects is condemned to its own rigid form of social structure.

In contrast, human societies have often retraced their steps and started on a new course. At the risk of repeating myself, I want to place special emphasis on this point because the ability of the human species to change, at will, its social and cultural patterns accounts in large part for its uniqueness among animal species.

Except for a few adaptations of minor importance, animals can change their lives only through exceedingly slow transformations of their genetic endowment, which require large numbers of generations to have significant effects. Moreover, genetic changes proceed almost irrevocably in a given direction. It was through innumerable minute progressive steps, all in the same direction, that the dog-sized species *Hyracotherium* (also known as *Eohippus*) evolved in some 50 million years to become the present-day horse.

In contrast, time and time again in the course of the human adventure, social and cultural patterns which had been in existence for centuries have been abandoned and replaced by other patterns, often with an opposite character. A classical example is provided by the Roman Empire, which at the time of Augustus 2,000 years ago appeared fixed in its course. Yet practically the whole Roman world changed from paganism to Christianity under the impact of religious and social reformers. The history of humankind is the history of similar changes— from the hunter-gatherer way of life to agriculture and pastoralism; from polytheism to monotheism or agnosticism; from the monarchy of divine right to the various forms of democracy; from classicism to romanticism in art and in literature.

The unique ability of the human species to change the course of its social evolution gives a deeper significance to the role of countercultures in history. Countercultures have been a constant manifestation of human life in all types of human societies. They constitute the experimental task forces which call attention to the need for changes and thereby help humankind to renew its social structures. In our own times, countercultures have been the public expressions of dissatisfaction with the purely materialistic values of technologic societies. Ever since the nineteenth century, it had been widely stated that capitalism was responsible for all the social problems of industrial societies. But countercultures have now made the Western world conscious of the fact that the essential failing of its societies is that they are not oriented toward real human values. Whether the structure is capitalistic or communistic, we base our choices and decisions on technologic means rather than on human ends; our criteria are power, efficiency of production, and quantity of consumption, rather than the quality of human life. By selecting as their targets the goals of consumer societies, the countercultures are taking a position which corresponds to that of the social forces which attacked the materialism of the Roman Empire and tried—with little success—to replace it by the more human values of early Christianity.

Many social changes have begun as violent revolutions which were defined better by what they wanted to destroy than by what they tried to create. In general, countercultures have taken a rather mild course, but they too have had a clearer vision of what they disliked than of what they wanted to create. The advocates of violent revolutions and of gentle countercultures have in common, however, the hope that human societies will be reborn after the great upheaval or the great awakening, even though it is difficult to predict in detail the forms they will take after the resurrection. In fact, utopias which are designed with too much precision and detail soon become prisons, as are the highly structured societies of social insects; the social formulas best adapted to the human species are open systems. To know the precise destination of the social process is not as important as to believe that social creation continues and that all men of good will can participate in the creative process.

From the dispersals of the Old Stone Age humankind acquired a taste and perhaps actually a need for change and adventure. The time may have now come for the Western world to take once more to the road, figuratively at least, not so much to escape from civilization as to give it a really new form. No one can predict what this form will be except that it must be compatible with the characteristics and needs which define the human species and which are essentially unchangeable. The future of humankind depends less on a new creation than on a resurrection with all the uncertainty and joyous hope associated with the resurrection of nature in the springtime.

V / On the Pleasures of Being Human

1 / The Diversity of Human Life

I wrote the last pages of this book in Aspen, Colorado, in the heart of the Rocky Mountains. Even though Aspen is well known to the American public, I must give a few facts concerning its geography and history for the sake of foreign readers, and also to help explain why my impressions of this enchanting place were distorted at times by memories of my youth.

Aspen has grown in a sinuous valley of a massive mountain range; sharp peaks of red sandstone emerge from a basis of Precambrian granite festooned with colorful lichens. As soon as the snow has disappeared from the valley and the slopes, the whole landscape is bedecked with a rich vegetation of bright wild flowers and of very tall grasses which sway under the most gentle breeze. The name Colorado truly fits this colorful jagged scenery.

Until the arrival of the white man, the region harbored only a small Indian population belonging to the Ute and Mowatavi-Watsui tribes, who spoke an Aztec type of language. These Indians left the region at the end of the summer and moved south to escape the snow. The Aspen area was therefore unspoiled wilderness when a few white men began to settle in it after 1859, apparently to trap beavers which were then abundant. Aspen was only a small mountain settlement known as Ute City in 1880, when silver was discovered in the mountains.

The region was then immediately invaded by white prospectors, and the town, which soon reached a population of 15,000, displayed the boisterous prosperity associated with silver and gold mining. In 1893, however, silver was demonetized by Congress and the mines were immediately abandoned. The Aspen population fell abruptly to a few hundred people, most of them of European farming background. They began to raise sheep and cattle, along with a few crops, moving their livestock to the south away from the snow at the end of summer. This practice is continued by the modern ranchers, but the animals are moved by truck instead of on foot.

During the past two decades tourism has become the chief industry of Aspen. The snow is abundant and beautifully dry throughout the winter—ideal for skiing; the bright sun adds to the enjoyment of skiers and tourists. During the summer the heat is never excessive despite the brightness of the sun. The showers which occur almost every afternoon keep the flowers bright and maintain an abundant flow of water in the lovely brooks and rivers that tumble from the snow-capped summits.

The natural advantages of Aspen have made it a paradise for tourists, and also an important center of culture. Concerts, spectacles, lectures, and a wide variety of symposiums and discussion groups enliven the intellectual life of the valley, especially during the summer months. Naturally, intellectual life is accompanied by innumerable receptions, cocktail parties, and other social manifestations. In addition to the ordinary tourists in search of pleasant climate and beautiful scenery, Aspen attracts students, professors, artists, and businessmen who come to assimilate some culture while in contact with nature, without having to forego the gossip to which they are accustomed in their urban circles. Many visitors engage publicly in various oriental practices of athletics and meditation, in the hope of relating more harmoniously to the cosmic forces which manifest themselves so dramatically everywhere in Colorado.

Aspen thus offers a rich fare to the senses and the mind. Yet I experienced a peculiar kind of uneasiness during the early part of my stay there, because I was constantly comparing what I saw and heard with the image of the Far West I had formed during my youth in France. When, as a little boy, I used to walk in search of adventure through the tame forests, pastures, and cultivated fields of the Ile de France, the exotic sonority of the very name Colorado evoked in my imagination tribes of heroic and picturesque Indians, as well as lonely cowboys, trappers, and prospectors. But instead of the Kit Carsons, Billy the Kids, and Buffalo Bills whose exploits had peopled my youthful dreams, I had to be satisfied in the Aspen area with urban people rigged in fanciful or careless "western" outfits unworthy of my noble image of the Far West. The human atmosphere was extremely friendly, but it appeared somewhat artificial in comparision with that of my youthful dreams. There were endless greetings of "Hi" and "Hello," as in the days when encountering another human being along a trail or a river was a notable experience, whereas in fact the encounter was at a supermarket or a restaurant, or on a much traveled and well-marked trail.

The costumes of most of the tourists had the kind of slovenliness and untidiness which curiously remains academic and bourgeois. The conversations also were largely academic and dealt chiefly with urban and international matters. They commonly referred to Plato or Machiavelli, but never to Indians or cowboys. They dealt with electronic or classical music but not with the magnificent thunder which reverberated every afternoon from valley to valley. They expressed concern about the fate of the whales and of the equatorial forests but not about that of the animals and plants of Colorado, even though many of these are threatened by tourists and motor cars. Thus the intellectual atmosphere of Aspen during my stay there in the summer illustrated a tendency which is almost universal in the modern world of culture—to replace the direct sensory perceptions of the world in which we live by preoccupations based on indirect knowledge about things far removed from our actual experience.

The excellent programs of the Aspen music festival symbolized this dissociation between cultural activities and the immediate environment. A particular program of chamber music I attended presented Bach, Beethoven, Schumann, and Hindemith. The Schumann lieder were based on romantic poems by Joseph Freiherr von Eichendorff about the sorceress Lorelei and her mysterious castle by the dark waters of the Rhine. These poems were famous a century and a half ago because they reflected the romantic mood of the period. But their brand of lyricism appeared to me unsuited to the luminous Colorado sky, among a public consisting largely of tanned half-naked young adults obviously more interested in the satisfactions of the body than in the preoccupations of the nineteenth-century soul. It is doubtful that von Eichendorff and Schumann would have composed their kind of poetry or music in the sparkling and shamelessly intoxicating atmosphere of the

Rocky Mountains in 1973. I could not help feeling that the dissociation between academic culture and the immediate environment contributes to the anemia of artistic creativity of our times.

On the the other hand, the simultaneous presence in Aspen of talents originating from all parts of the world was exciting evidence that our civilization has made progress toward a global view of humankind. The universality of humanistic concerns—from religion to the arts and to technology—manifests itself in the form of meetings held wherever experts can assemble under pleasant conditions—in the Aspen valley or on Lake Como, in the Polynesian Islands or along the Mediterranean shores, in mountain or desert. At these meetings a fundamental unity appears even in the menus, with much the same cocktails, chicken or filet mignon, hearts of palm or avocado pear, Scotch or brandy. Internationalism is now reaching simultaneously both the mind and the stomach.

The trend toward internationalization of human life follows from the biological and psychological unity reaching back to the Old Stone Age. Everywhere on the prehistoric scene *Homo sapiens sapiens* presents himself with a skeleton, a cranium, artifacts, and occupations which can readily be assimilated to our own. We can use his tools and his weapons. We admire his workmanship as well as the esthetic quality of his engravings, paintings, and sculptures. We are beginning to understand the notations on bone and ivory by which he recorded the rhythms of life and the seasons, and probably also the phases of the moon. We are still moved by the mysteries he tried to convey through his arts and rites and by the care he took of his dead. In brief, we feel that we understand Old Stone Age life almost as well as that of our own period, for the simple reason that the body and mind of the human species have hardly changed in the past fifty millennia. This fundamental stability of human nature is so important for the understanding of modern life that I shall illustrate it once more with an example from the history of Aspen.

The first European settlers in the Rocky Mountains came from a wide variety of trades and social classes—farmers and artisans, administrators and merchants, men of letters and scientists, criminals escaping from justice and men of God bent on saving souls. Although they had behind them a long-lasting and complex civilization and had been brought up in highly humanized environments, they acquired rapidly the techniques necessary for survival in the wilderness. Through an atavism that reverted to the Old Stone Age, they learned to deal with wild animals and to engage in life styles similar to those of the prehistoric hunter-gatherers. Then, one or two generations later, the descendants of these trappers, prospectors, and *coureurs des bois* went back to civilized life and became tradesmen, businessmen, or university professors. It is the legendary life of the pioneers that the Aspen tourists now try to act out, with behavior patterns and clothing styles that would be somewhat ridiculous if they did not acquire dignity from the relation of that life to the beginnings of the human race.

While all human beings operate from a common biological basis, their responses to the environment, their behavior patterns, and their search for satisfactions of the body and the mind always have a local character because the spirit of the person has to express itself through the spirit of the place. This explains why national characteristics retain their distinctiveness in the modern world, despite the diffusion of cosmopolitan technology.

In his acceptance speech of the 1972 Nobel Prize for Literature Aleksandr I. Solzhenitsyn stated: "The disappearance of nationalities would impoverish us no less than if all people were to become identical, with the same personality and the same face. Nationalities are the wealth of humanity, they are its crystallized personalities; even the smallest among them has its own special colors, hides within itself a particular facet of God's design." [1] In the first chapter of his novel *August 1914* Solzhenitsyn presents Russian nationalism not as an idea but as an affective reaction, deeply rooted in the subconscious. Russianness has something to do with the soil, and everything to do with the people, their beliefs, and their language. It is unrelated to political ideologies and is as natural as a tree, deriving its nourishment from the subsoil but rising toward the heavens. For Solzhenitsyn, Russia is not a place on the map but an image shaped by the Russians—much as, for George Orwell, England was an attitude formulated by Englishmen.

The formative influence of the environment is commonly so profound that a territory once occupied by a given people may be lost without this loss causing the breakdown of national identity. Such persistence of identity is well recognized in the case of the Jews, but it has also been observed with many other people. For example, the Yaquis, the Navajos, and the Cherokees were displaced from all or part of their territories yet survived as peoples. National individualism can also survive political domination. The Irishmen of the present Republic of

Ireland feel a continuity with the Irishmen of more than 1,000 years ago; they have steadfastly rejected identification with Englishmen even though their country was part of Great Britain for a long period of time. Tragic experiences commonly help people to retain their collective identity under a wide range of physical and sociocultural environments.[2] Except in a limited biological sense, in fact, we are shaped less by our environment and our past experiences than by the image we create of the past.

Aleksandr Blok, one of the Russian symbolists of the early twentieth century, tried in *The Collapse of Humanism* to differentiate between civilization and culture. For him, culture is the "musical" reality that underlies everything, the sense of complete harmony between spirit and flesh, man and nature—a truly elemental force. "Great is our elemental memory; the musical sounds of our cruel nature have rung in the ears of Gogol, Tolstoy, Dostoevsky."[3] In contrast, civilization emphasizes material possessions, calendar time, the growth of specialties, and other cosmopolitan characteristics. Vague as it is, this distinction between culture and civilization helps to explain why international technology has not yet destroyed national individualism, and it points to the mechanisms which have made the national spirit such a powerful creative force. Intellectuals and artists, as well as peasants, best contribute to the universality of humankind by giving form to the unique resources of their immediate environment. The music composed by Bach, Beethoven, Schumann, and Hindemith continues to have esthetic and emotional value for modern people. But while its performance in Aspen was a source of enjoyment, it did not add to the wealth of humankind because it did not express the genius of the twentieth century in the Rocky Mountains.

All cultural forms tend to repeat themselves except for technical improvements, or they try to remain creative by orienting themselves toward abstractions further and further removed from the concrete world. Fortunately, there persists in human nature a desire for sensory contact with the immediate environment. Whereas cultures soon suffer from anemia in the world of abstractions, they rapidly acquire a new vigor when they re-establish direct contact with the concrete world through the senses. History is replete with examples of civilizations which have known the boredom of academic rigidity and the chill of abstractedness during certain phases of their evolution and then have recovered a creative exuberance when conditions have compelled them to be in direct contact with nature. The very early part of the medieval era, far from being a Dark Age, as used to be thought, was in reality a rich period of renewal. The breakdown of the Roman Empire compelled the different regions of Europe to live almost in isolation. By having to derive food and entertainment from local resources, each region was given the opportunity to develop its own styles adapted to its environment. Similarly, the Vikings and other Nordic invaders, after having adopted the principles of Christianity and of Roman civilization, created life styles and artistic forms adapted to the local conditions of the places where they settled. The endless occurrence of upheavals in the early Middle Ages renders even more remarkable the vitality of the period, from which emerged styles peculiar to each region of Europe and even to each locality. Agricultural and technological practices, forms of art and of literature, ceremonies and saints, as well as foods and beverages, were but a few among the countless expressions of spirit of place which survive even today in the prodigious diversity of European cultures.[4]

The most important achievements of humankind have a universal character in the sense that they express values common to the whole human species, but each has special forms conditioned by the time and place of its emergence. These special forms contribute to the endless renewal of civilizations. The Greek fable tells that the hero Antaeus lost his strength as soon as both his feet were off the ground. The same can be said of civilizations, which become weaker as they lose direct contact with reality. An even older myth is the myth of resurrection. We now readily accept the thought that all civilizations are mortal, but we know also that they can be reborn by feeding on reality through the senses. The plant dies, but its seed becomes a plant again by incorporating the substance of the earth.

2 / Pluralism and World Order

Persons, cities, and nations acquire in the course of their development characteristics which give to each of them an increasingly distinctive personality. The trend toward individualization and therefore toward pluralism can be seen at all levels of living processes. But there is also in life an opposite trend toward the unification of organic and social structures. For humankind, the ultimate expressions of these two trends are on the one hand the cult of personality and on the other hand the efforts to establish a world order integrating all nations into a single political body. There seems at first sight to be a contradiction between the two tendencies, one toward increasing diversification and the other toward the creation of a higher order of unity out of this very diversity. But in fact, both are essential not only to the social evolution of the human species but also to the purely biological evolution of all living things.

While it would be naive to assume that human problems are entirely explainable in terms of classical biology, some lessons can perhaps be derived from the intriguing analogies that exist between purely biological phenomena and the evolution of human societies. I shall therefore mention here a few facts which illustrate how, in nature, organic life evolved progressively from a low level of unity toward a higher level of integration through the intermediary of pluralistic diversification. This brief excursion into the domain of theoretical biology will also help to sharpen the qualitative distinction between biological evolution and the social evolution of humankind, which can be simply stated as follows: In ordinary biological systems, the progression from a low to a more complex level of unity occurs through a series of unconscious evolutionary phenomena, whereas in human life it involves conscious decisions taken to reach goals which appear desirable even if they are in conflict with biological necessities. The operations of free will constantly disturb the deterministic course of human affairs.

All living things present great similarities in their chemical composition and in their fundamental physiological functions. It is therefore almost certain that they all derive from a common origin. The nucleic acids, amino acids, and other complex chemical structures which are essential components of all living organisms could not possibly have had, as they do, the same fundamental composition in such organisms as an amoeba, an asparagus, an oak, a giraffe, and a man, if they had not emerged from a common source. Chance could not possibly account either for the fact that similar kinds of sugars, enzymes, and vitamins are essential for the elaboration and operation of structures as different as the membranes of microbes and the cells of the human brain.

For lack of another explanation, it seems reasonable to assume that vital processes originated from relatively simple chemical reactions in some kind of very primitive structure—the first living form. *A priori,* this hypothetical living form could have grown in size without changing its characteristics, thus becoming progressively a gigantic creature which would have constituted the unique form of life. But what really happened is of course very different. Instead of simply growing and reproducing itself, the hypothetical living substance progressively underwent modifications and gave rise to innumerable forms increasingly different from one another. The outcome of this evolutionary process is that life is now represented on earth by countless species, all of which are built according to the same fundamental pattern, because they evolved from a common origin some three billion years ago.

The theory that all living forms have a common origin has received experimental confirmation from a spectacular laboratory tour de force. By associating the cells of different species under very special laboratory conditions, one can make them exchange their genes and thus generate new hereditary systems. Such genetic recombinations have produced artificial hybird cells which are viable, yet are derived from two very different species. New artificial plants have thus been produced through the agency of genetic recombination. The genetic structures associated in such hybrid forms had not had any contact during the immense periods of time that elapsed while they were evolving into different species. Yet the cells had retained the biological memory of their common origin—a memory that enables them to unite when placed under the proper conditions.

While retaining a fundamental unity of structure, the various species of animals and plants continue to evolve and thereby to become more and more different one from another. In nature, however, even widely different species can form associations to create complex social organisms, through mechanisms which do not involve genetic recombination. Such associations are designated by the word symbiosis. The symbiotic association involves a complex integration of certain structures and activities, resulting in the formation of what might be called a new social organism. But this integration can be readily disturbed, as is illustrated by the case of the lichens. The numerous varieties of lichens—plants which commonly occur on rocks and tree trunks—are symbiotic associations of two very different kinds of microbes—algae and molds. These symbiotic associations possess remarkable properties which enable them to multiply where nutritional resources are extremely poor and to survive drought, cold, and heat. Air pollution, however, rapidly disorganizes the integration between algae and molds, thus destroying the lichens. For this reason lichens are extremely scarce in polluted cities such as New York, London, Paris, Tokyo, to the extent that their presence or absence is one of the most sensitive indexes of air pollution.

I have selected the example of lichens and their disintegration because of its simplicity. It illustrates how two different species can become associated to produce a new socially integrated organism, the characteristics of which transcend those of its component parts, and also how the continued success of this integration depends upon external circumstances. There exist in nature many examples of much more complex symbiotic associations. The word ecosystem designates an association between a certain physical environment and all the living things it harbors. A forest, a meadow, a lake, or even a city constitutes an ecosystem in which all parts interplay and influence each other, reaching a state of equilibrium which gives its individuality to the system—the genius of

184

the place. As with the lichens, however, any disturbance of this "ecological" equilibrium can have disastrous consequences, as is presently evidenced by the environmental crisis.

While the tendency to form complex ecosystems is universal and is manifested also in the creation of human societies, there exists in human nature, as mentioned earlier, an opposite drive toward biological and intellectual individualization. This drive is probably essential to the development of civilizations. "To compose our character is our duty," Montaigne wrote. For each of us, the shaping of our character can be, in his words, "our great and glorious masterpiece." [5] But this can be achieved only by living in a manner suited to the special conditions of our existence. Ideally, human beings should therefore strive for an intimate adaptation between their own unique organic and psychological endowment and the circumstances of their individual lives, which are necessarily unprecedented. Since such adaptive interplays differ from one person to another and according to the situation, diversity is an inevitable consequence of life and especially of civilized life.

As in the biological domain, however, highly individualized persons can become socially integrated and thus generate new social organisms whose characteristics transcend their own. The existence of nations constitutes the most spectacular and the most subtle illustration of the fact that social organisms can generate a new unity out of the diversity of their individual components. All great nations are made up of a multiplicity of races, professions, philosophical and religious beliefs, as well as of persons living at different economic levels. And yet it is obvious that each nation has its own personality. The names United States, Canada, England, France, Germany, Mexico, Brazil evoke certain geographic and demographic traits which are peculiar and distinctive for each one of these countries, and also certain habits, attitudes, and ways of life, resulting from the integration of these traits with historical forces. From this integration there emerges a new unity of a higher order than that of the constituent parts; its expression is the national genius.[6]

The two opposite trends, one toward diversification and the other toward unification, have both been creative forces in human history. Human beings have become progresssively differentiated into a multiplicity of biological races and social groups in the course of their migrations, but they have always been involved in exchanges which have enabled them to create lasting associations and to move slowly toward some form of social unification. From family to clan, to tribe, to nation, and to federation of national states, one can recognize since prehistoric times a halting but constant trend toward a world federation.

The evidence that trade routes existed across Europe during the Old Stone Age and that the first cities of the Near East were centers of exchange shows that the process of unification began many thousand years ago, long before the historical period, even before the development of agriculture.[7] A marked acceleration of social change occurred after the agricultural revolution, and the emergence of city states, nations, and empires soon followed. Since, as far as can be judged, the biological and psychological structure of humankind had reached its present state much earlier, this acceleration of social change can hardly be explained by the emergence of new mental qualities in the human species. The French anthropologist Claude Lévi-Strauss has suggested that it was caused by increased contacts between different human cultures. In a report prepared for UNESCO in 1952, he expressed the opinion that individual cultures evolve only very slowly as long as they remain isolated. But their social evolution proceeds rapidly when they exchange information and goods with the external world, even though each culture retains its originality.[8] To this hypothesis it might be added that the interplay between cultures commonly takes the form of associations analogous to biological symbiosis—that is, resulting in the creation of new social organisms with characteristics transcending those of the component parts. If this analogy is at all justified, one can postulate that symbiosis between different human cultures will enhance their differentiation and thus increase the diversity of social systems.

The trend toward diversification of social systems may appear undesirable from the international point of view since it goes counter to the political and economic unification of the world. But there is evidence that, as in purely biological systems, diversity encourages the emergence of organs of communications which generate a new and higher level of unity. Each nation, region, or city acquires uniqueness—its peculiar genius—through the interplay between the traditional culture of its population, the resources and other characteristics of its natural environment, and the foreign cultures to which it is exposed. This process of differentiation will probably continue despite the homogenizing influences of technology, commerce, and travel. Today as in the past, various countries continue to specialize in certain types of industrial and economic activities and also to emphasize the unique aspects of their political, philosophic, and religious attitudes. Democracy is different in England from what it is in France, in Canada from what it is in the United States. The Catholic religion is not the same in Amsterdam as in Naples, in Mexico City as in Boston.

In biological systems, the functional specialization of the different parts commonly results in the emergence of mechanisms for their integration. For example, the various parts of the human body are extremely specialized but they behave nevertheless as a unit because they are integrated through the agency of hormones and of the nervous system. A similar process of integration can be recognized among the various nations of the world. It began long ago with personal contacts and mechanisms for the exchange of goods; it now involves complex international organizations which operate simultaneously in all the nations of the globe.

The problems posed by wars, epidemics, pollution, depletion of natural resources, labor conflicts, monetary crises have evoked a series of social responses which have generated complex systems of communication between the different countries. The World Health Organization, the World Meteorological Organization, the World Bank were among the first international structures expressing this type of response; the international structure which was set up after the Stockholm Conference on the Human Environment in 1972 is the most recent of these responses to a global problem. And the process will certainly continue. The more the nations and regions of the world continue to differentiate—through their economic activities, their social structures, and their ways of life—the more they will need international mechanisms to integrate their activities, especially in view of the fact that modern technology makes them interdependent with regard to natural resources. Nations will thus achieve together a functional, organic unity while retaining their diversity.

This continued process of integration does not imply that the world of the future will be completely uniform and that life will become boringly standardized. The only aspects of culture that become really cosmopolitan are those which are most readily influenced by the mass media and which deal with a few secular aspects of life. It is probable that the natural sciences and the great technologies derived from them also will bring about some standardization. But each particular culture will continue to search in its own way for certain forms of truth and for a quality of life compatible with its traditional genius. There are many ways of organizing a given number of trees within a given area, and thus of creating either a Japanese park or an English park or a French park, each with its own symbolic meanings.

In all living systems, including human societies, integration emerges along with differentiation. The world order which will eventually result from the integration of the different social structures will constitute a higher form of unity, compatible with the pluralism of tastes, ideologies, and spiritual aspirations.

3 / Adventure and Fantasy

Our greatest blessings, says Socrates in Plato's dialogue *Phaedrus,* come to us by way of madness. Plato did not mean, of course, that being insane is better than being sane; he rather pleaded through Socrates's voice for the freedom and fantasy that enable us to transcend conventional life. Spokesman of rationalism and of social order as he was, he acknowledged nevertheless that original creations come chiefly from the subconscious forces designated in his time by the word *mania* and associated in Greek mythology with Dionysus, Aphrodite, Eros, and the Muses. Dionysus was also Eleutherios, "the liberator," who enabled the human spirit to escape from its corporal bondage through *ekstasis*—a word which meant literally "being out of one's self" and which denoted a profound alteration of personality.

The cult of Apollo added conscious discipline to the subconscious acts of creation inspired by the mania; in the course of centuries, reason has thus come to play an increasing role in the conduct of human affairs. The exact sciences and the technologies derived from them are now the most spectacular expressions of this rationalism. But while they have given a more precise form to understanding and have enormously increased material wealth, they have somewhat inhibited the expressions of human fantasy. And yet it is still from fantasy that human beings derive their richest satisfactions.

The human species is naturally governed biologically by the same laws as other forms of life; furthermore, the stability of its genetic constitution greatly limits its chances for progress through Darwinian evolution. But human beings are endowed with a spirit of adventure and fantasy that enables them to move into many different environments and to create innumerable social structures. Creativity is a consequence of the desire for adventure. The longing for new information, the urge for new undertakings, and the many forms of restlessness are the equivalents of the ancient call of the road. But while leaving camp now and then may be a response to a fundamental psychological need, taking to the road has its own rewards. As Cervantes wrote, the road is better than the inn. Or, according to a Spanish proverb, the road that leads to Paradise is itself Paradise. Simple ambition and the hope of achieving power or wealth cannot alone explain why, from the first prehistoric explorations to those of the astronauts, human beings have always been attracted by unkown places. Mystery seems to create for them a magnetic field in which they willingly get caught because they derive satisfactions from the experience.

Travel provides the occasion for many elementary pleasures and for intoxicating thoughts. Going by foot, by horse, by ship, or even by automobile engages the body in a rhythm of movement which relates it organically to the landscape through all the senses. Movement also engages the mind by generating the hope of the unexpected—a new type of landscape, an alluring encounter, the discovery of places or people suggesting the possibility of starting life anew. Displacements, accidental or voluntary, increase knowledge of the physical and social world; they enlarge the scope for adventure by providing a more acute awareness of human potentialities and more options from which to choose.

The true traveler, the one who leaves camp not out of necessity but for the sheer pleasure of traveling, derives part at least of his motivation from the very human trait of impatience with the status quo. Human beings are prone to grow weary even of conditions that at first had appeared agreeable and desirable. A dog or a cat will happily eat the same food day after day, but human beings ask for variety in food as they do for variety in fashions, automobiles, and art styles. This lassitude toward the status quo may contribute to the fact that civilizations commonly deteriorate when they cannot move forward. The sense of panic created by the previously mentioned book, *The Limits to Growth,* has its origin in the general belief that if a ceiling is placed on quantitative technological development stagnation will engulf us—fortunately an erroneous belief.[9] From a more positive point of view, taking to the road can be creative by giving a wider field to the exercise of free will.

Most human beings have some latitude in selecting the conditions under which they live. They move into environments or adopt customs that appeal to them, rather than merely accepting the conditions under which they have been placed by the accidents of life. Millions of Europeans migrated to America, and many, I being one of them, made the move out of choice and not from necessity. To a large extent, it is therefore as a consequence of their own decisions that immigrants have been shaped by the American environment. In the original French text of this book I found it easy to express my conviction that environmental conditioning usually depends in part on willful actions: *"Puisqu'une grande partie de la vie se passe sous le signe de l'engagement volontaire, l'évènement façonne moins la personne qu'il ne répond à son appel."* But I have not been able to put succinctly into English this statement which is the central theme of my book. An approximate translation might be: Since a large part of our life is an expression of our own choices and decisions, surroundings and events are in the final analysis of less importance in shaping us than is the fact that we first choose to expose ourselves to their influence.

Just as human beings have latitude in selecting the channels in which they engage their lives and therefore the conditions that will influence their subsequent development, so they have freedom in cultivating one or another among the multiple and often contradictory aspects of their psychological attributes. Egoism or altruism, violence or tenderness, the spirit of domination or that of conciliation, the desire to take to the road or to settle in a sedentary life are but a few of the multiple options among which each human being has to select. For most of us, it is all but impossible to know ourselves by a pure process of reflection or meditation; we know who we are only by examining the choices we make and our responses to new conditions.

Throughout history and prehistory humankind also has had to choose repeatedly between opposite tendencies which seem to be inherent in human nature. For example, love and aggressiveness have roots which are equally deep in the evolutionary past. Of the two, love is obviously generally preferable, but there are many situations in which aggressiveness is necessary for the survival of the person or the group. It is through a consecutive series of such choices that humankind has converted certain elements of its primitive, untamed past into the complex social structures which now constitute the different civilizations. The kinds of attitudes and actions among which humankind must select have naturally varied in the course of time, but there are certain dilemmas which appear eternal. The apple offered to Eve remains a valid symbol for the dangers which inevitably result from an increase in knowledge. The serpent no longer worries us, but temptation still symbolizes the painful uncertainty of having to choose between the status quo with the comfort it provides, and adventure with the enrichment of life it makes possible but also the dangers it entails. And yet choose we must. To be human implies the necessity and also the possibility of choosing among options of which each has its merits—between cooperation and individualistic action, between theoretical knowledge and practical application, between spiritual values symbolized by contemplation and mastery over nature made possible by technology.

The possibility of choosing is an essential factor of the human condition, but it is inoperative without a vision of the future. The ability to imagine the future, to "invent it," in Dennis Gabor's felicitous phrase,[10] is a prerogative of human beings. Through anticipation they can modify the course of natural events because they are capable of influencing consciously the unconscious determinism of nature. The hunter learns about the behavior of animals so as to be able to anticipate their positions and thus to attack them more effectively; the farmer learns from experience how to modify the soil, the flora, and the fauna in order to grow a special kind of crop. The technologist visualizes a certain type of industry before extracting from the earth the natural resources he needs. In practically all human activities, the anticipation of the future thus precedes interventions into nature.

Social structures likewise reflect the hopes of planners. The Declaration of Independence states certain social principles which were not self-evident, despite the affirmation of its signers, and certainly are not universally accepted but which corresponded to the social patterns that the American colonists had in mind for the ways of life they were creating in the New World. Similarly, the French society based on the "Declaration of the Rights of Man and the Citizen" was not the expression of ineluctable laws governing human nature but only of a view of the human condition at a particular moment of history. If a similar declaration had been written in Athens 2,500 years ago, it would have been different from the one written in France in 1789 and would, for example, have accepted the practice of slavery. If the French Declaration were written today, it probably would include other rights of man, including the right to a healthy and pleasant environment.

Choices intervene also in esthetic questions and they influence not only the plastic arts, the literary forms, the proprieties of language, but also the structure of landscapes. Only after the eighteenth century did Europeans realize that there was beauty in the wildness of the Alps. Only then also did their admiration of French classical gardens yield to the craze for the seminatural English parks. And the present cult of untamed wilderness was still to come. The ocean also was long ignored by artists. Until the end of the sixteenth century England was chiefly agricultural; she discovered the sea as a result of her conflict with Spain and from then on found in it adventure, wealth, and inspiration.

The ideal of feminine beauty has constantly changed throughout prehistory and history. Two extreme expressions are the obese Venuses of the Old Stone Age and the slender Pre-Raphaelite heroines of the nineteenth century. The contrast between these two styles certainly reflects a change in taste rather than in feminine anatomy. In fact, tastes in this regard seem to have changed even in the course of the Old Stone Age. Whereas statues of women celebrated fleshy opulence and sexuality in the period of the Venus of Willendorf, they acquired a more ascetic character somewhat later in the Stone Age.[11] Esthetic ideals with regard to the feminine body continue to change. A pronounced goiter enhanced the charm of women in certain parts of Europe during the seventeenth century; the nineteenth century romantic heroines had to be pale and emaciated; and at present women of all ages manipulate their diets and their bodies in an effort to look like Diana the huntress.

For several centuries now, material wealth has been the social ideal of the Western world; progress is identified with scientific technology and with the use of machines in all human activities. During the same period, Oriental civilizations have been focused on a better knowledge of human nature—its artistic, philosophic, and religious expressions as well as the anatomic and physiological control of the body. The fact that Western people are becoming interested in such Oriental practices as acupuncture and yoga shows that even scientific trends are influenced by choices which have their origin, not in the internal logic of science, but in the longing for intellectual adventure.

One aspect which is common to all civilizations, whether primitive or technologic, is that humankind has been able to use its physiological and psychological needs to transcend its animality. Thus, sexuality evolved very early into eroticism; the medieval courts of love gave rise to a desexualized cult of woman, with its own forms of literature and art. Even the need to eat generated its own form of art, as exemplified by Anthelme Brillat-Savarin's *Physiology of Taste* (1825) or the *haute cuisine* of Mandarin China. This conversion of animality into purely human interests which represent some of the most interesting philosophic and artistic achievements has proceeded through peculiar phases in the course of its evolution within the human brain. The bulls depicted on the walls of Altmira, Lascaux, and other prehistoric caves certainly had a symbolic meaning for the Cro-Magnon hunters. But bulls also appear in the art of ancient Crete—and in Picasso's painting *Guernica*—with a significance certainly different from that of the ones in the Old Stone Age paintings.

Almost all aspects of nature have thus been converted by culture into abstract values which link human beings to cosmic forces. The flowering of springtime, the sumptuous vegetation of summer, the brown melancholy foliage of late autumn, the white luminosity of snowscapes are natural phenomena which acquire a philosophic and emotional significance only through humankind's creative imagination. The spiritual and emotive quality of nature does not reside in brute facts but rather originates from the transmutation of these facts into new values by human imagination and fantasy.

4 / *Joie de Vivre* and Happiness

The signers of the Declaration of Independence affirmed that the pursuit of happiness was a natural right of man. But the wise men of Buddhism and Christianity have long taught that one cannot find happiness by seeking it; one can only experience it by being involved in something else. One of the causes of perennial dissatisfaction in the modern world may be the frantic pursuit of happiness.

Like other human beings, I have encountered happiness while submerging myself in the external world through my body and my imagination. I experience it especially early in the springtime when organic life awakens and everything once more appears possible. The harbingers of spring and the promises of Easter mean more to me than the plenitude of summer. Bubbling water springing up from the earth after the thaw evokes for me the creation of the world; the swelling of buds confirms my faith in an eternal renewal of life. I experience happiness also at daybreak when the air is still unsullied and the day innocent. Whether the atmosphere is transparent and gives a feeling of assurance or whether a mist shrouds the world in mystery, it is good to hear life start anew with the first sounds of nature and of the street. In the early morning, I too hear the call of the road.

I believe I have recognized certain ingredients of

happiness in the descriptions by anthropologists of what human life may have been during the Old Stone Age—on the plateaus of East Africa where humankind seems to have emerged, on the cliffs bordering the valleys of the Vézère and the Dordogne which sheltered Cro-Magnon man, on the slopes of the Zagros mountains from which came the people who created the first civilization in Mesopotamia. I imagine all these early representatives of humankind in a somewhat harsh but invigorating climate, before a vast horizon reached through an uncluttered and varied landscape of meadows, rivers, lakes, and clumps of trees. From my readings about bucolic life in later periods, I add in imagination to these primitive sceneries the straight furrows cut by plowshares and the call of hunting horns.

Other ingredients of happiness appear to me in the pictures where artists have shown scenes of collective life at all periods of history. In general, the groups are rather small and fairly homogeneous, but if the picture shows a large and motley crowd, the people seem to achieve unity through a special event or a common cause. Whether the occasion be rustic as in a Breughel painting or urban as in scenes of the Italian Renaissance, the gathering takes place against a varied backdrop with dimensions to the measure of man, neither cluttered nor too empty.

The scenes in which I can detect the elements of happiness resemble those described in innumerable fairy tales and legends. It would seem that humankind has always had a fairly clear notion of what life in Arcadia would have been, for the atmospheres which evoke happiness have much similarity in the mythologies and artistic creations of all peoples. Like our predecessors, we find it difficult to imagine happiness in the midst of a desert, in the depths of a dark forest, or in the over-crowded and polluted areas of a dull urban ag-glomeration. Rather, we situate happy life in an oasis, beside a body of water, on the edge of a woodland, in the open spaces of a village or city, or in the bosom of the family. Empirically and to a large extent unconsciously, the various forms of civilization have tried to reproduce, through mythology, literature, and the arts, atmospheres similar to those in which humankind achieved its biological and social identity. One can postulate that many primitive human beings did experience a kind of organic happiness in the environments where they evolved and to which they were biologically adapted.

The most elementary form of happiness is to feel life flowing through one's own body in an harmonious relationship to the rest of the world—the simple bio-logical *joie de vivre*. One can perceive it in the play-fulness of a kitten, a puppy, a lamb, or a foal in springtime and in the sense of physical well-being displayed by a cat stretching itself in the sun or by the hearth. It was prob-ably the same kind of organic *joie de vivre* that primitive human beings expressed through the exuberant or-namentation of their tools and weapons or by the erotic jests they depicted on the walls of some prehistoric caves.

Even today, one can recognize the evidence of a simple *joie de vivre* among primitive populations, and even among the economically depressed social classes of industrial countries. There are admirable photographs illustrating the carefree enjoyment of living experienced by so-called primitive people as they play with their chil-dren, chat with their companions, or participate in some tribal festivity. As Barbara Ward stated in a recent interview, "You can't live in Africa, where I lived for eight years, without seeing the enormous amount of sheer unadulterated fun which comes from being with your neighbors, sitting around, talking, dancing, singing— being."[12] I have in mind a film showing the expression of bliss in the smile and in the eyes of an Australian aboriginal child as she anticipated eating the fat, delectable larva she had just extracted from under the bark of a tree.

The human species has expressed *joie de vivre* through a great variety of rites and celebrations which have their origin not in its biological nature but in its fantasy—this wonderful attribute that enables us to transmute the world of matter into spiritual values and into creations without practical utility but giving significance to life. Excuses for collective ceremonies arise from birth, death, marriage; the initiation into the activities of adult life; commemorations of mythic, religious, or historical events. In one form or another, such ceremonies began at the dawn of humanity and they have taken many different forms according to the period and the region. If Jesus had lived in eastern Asia, rice and saké would have been used instead of bread and wine in the symbols associated with his ministry, but the external forms of a rite have only local interest. What matters is that all rites are the embodiment of human needs which must be universal, since their manifestations can be detected from the beginnings of prehistory into our own time. Rites and celebrations facilitate the integration of the social group by linking its past to its present and its future.

Probably it is through their monuments, their

ceremonies, and their rites that societies best express their ideals and reveal what they would like to become. If the urban centers of new modern agglomerations are commonly so depressing despite their wealth, it is because nothing in them transcends the trivial satisfactions of material existence. Appliance shops and clothing stores now occupy the central area which used to be the site of a shrine or church, of the opera house or townhall; instead of monuments to saints or heroes, neon advertisements catch the eye; the noise of automobile traffic drowns human voices and the sound of bells. The great deficiency of modern life is that it satisfies only material needs, without providing an equivalent for our times of the cathedrals, the banners, the bells, even of the hunting horns, which have for so long given expression to the human dreams of a richer life.

In the Western world, for nearly 2,000 years, the most important aspects of personal, social, and mystical life have been symbolized by the rites and celebrations of Judeo-Christianity. Holy bread, wine, banners, candles linked humankind organically to the satisfactions provided by the earth. The language of the liturgy, its songs, costumes, and gestures symbolized the place of humankind in the cosmos and in history. Likewise, all other religions have rites and ceremonies which make use of sensory experiences to transmute the small events of daily life into values which facilitate the integration of the social group and which embody concepts of humankind's place in the order of things.[13]

Festivals have also long given a sensory expression to such abstract concepts as patriotism or loyalty to a cause. With flags, uniforms, drums, martial music or simple songs, even with stereotyped academic speeches, numerous societies have succeeded in converting the necessity of living in a certain place at a certain time into a national ethos that operated as a creative force. It may be impossible to define a nation in scientific terms, but one must be deaf and blind not to recognize that rich human values have come from belonging to a given nation and that these values have given particular national flavors to the *joie de vivre*.

Both the religious and the national sense have become much weaker during the second half of the twentieth century and they have not yet been replaced by equivalent spiritual forces. Modern society is undergoing progressive disintegration; it has hardly begun to reform itself, still less to search for a really new form. The countercultures are readily defined by what they reject—the consumer society, colonialism in all its forms, economic and military nationalism—but none has yet formulated a positive doctrine that could serve as a new cohesive social force. They have failed in this task because, despite their brave claims, their proponents have not found a way to introduce new enriching experiences into their own lives.

Festivals have lapsed into increasing triteness as they have lost their religious or national significance. As long as weekends and holidays are only occasions for loafing and amusements, they will not contribute to social integration as did the symbolic festivities which were once meaningful to the community as a whole. And yet the fact that large crowds gather spontaneously on the slightest pretext suggests the persistence of a profound need for the ceremonial forms of collective life. In our own times, it has taken natural catastrophes, revolutions, wars, or protests against war to re-create an appearance of social unity, and then only for a short time. But there cannot be any true integration without a system of values accepted by the majority of the social group and generating a collective *joie de vivre*.

The disarray and disenchantment which plague our times show that physical comfort, abundance of goods, and control of disease are not sufficient to bring about either individual happiness or harmonious social relationships. As I have suggested, a more direct experience of our relationships to nature and to the cosmos would not only enrich our lives but might become a substitute for the religious sense. The psychological need for belonging to a community might give a modern form to the tribal spirit which has so long been basic to the social structure of humankind and is the historical justification of nationalism. Individually and collectively we would do well to cultivate in our relationships with the external world the subtle quality expressed by the Arabic word *baraka*—the sense of blessedness that attaches itself to buildings or objects after years of loving use.[14] We need also to give freer rein to fantasy—this marvelous human faculty which enriches and embellishes reality by acts of imagination.

Happiness is contagious and for this reason its expression is a service and almost a duty. As some Buddhists say, "Only happy people can make a happy world." The most useful citizens are not necessarily those who increase production and knowledge but rather those who generate *joie de vivre* around them. Optimism and cheerful spirits are indispensable to the mental health of technological societies. The most valuable personal attribute as well as the most useful contribution to collective happiness is the spontaneity of the smile.

On the Pleasures of Being Human

More than ever, in Rabelais's words, *"Rire est le propre de l'homme* [to laugh is proper to the man]." The best protection against fate is to face life with a smile. I am using the word smile with the meaning it has in the French phrase *"prendre la vie avec le sourire"*—not the stereotyped smile of the airplane hostess or toothpaste advertisements, but the tolerant smile that comes from an amused awareness of the mutability of things.

Certain situations are naturally more conducive than others to the *joie de vivre*. From a purely biological point of view, the best environment for the human species may have been the pre-industrial, pre-agricultural world based on the hunting-gathering ways of life. This may be the biological environment to which we are still best adapted, the biological paradise we have lost. But human beings have progressively created for themselves another kind of life out of values found in the arts and sciences, in rites and celebrations, in the search for the absolute. Like Ulysses, they are attracted by the ease of life in the land of the Lotus-Eaters but in general they continue to prefer adventure with all the efforts, sufferings, and risks it entails. Out of choice many human beings devote a large part of their lives to arduous tasks which are physically uncomfortable and even painful, yet from which they expect the nonmaterial satisfactions they regard as the only worthwhile form of happiness. In many circumstances, the realization of an ideal, even though difficult and painful, is the essential ingredient of happiness. There is no obvious explanation for the fact that human beings deliberately place themselves in situations which require strenuous efforts and cause suffering. Similarly, there is no obvious answer to the question that Leonardo da Vinci asked himself in his notebook: "*Leonardo, perché tanto peni*?" Why indeed take so much trouble when physical pleasure can readily be found in animality?

To understand the human species as part of the animal kingdom requires only knowledge of its genetic structure and organic reactions to environmental stimuli. But to understand its humanity, it is essential to know why so much of human life is devoted to tasks that have no immediate utilitarian application—the arts, the sciences, the ceremonies, the innumerable forms of self-sacrifice. Human beings derive their most profound satisfactions— true happiness—from those activities which are the furthest removed from animality. When the German philosopher Benno Erdmann was asked to express his views concerning modern trends in the sciences of man, he sadly replied, "In my youth we used to ask ourselves anxiously: what is man? Today scientists seem to be satisfied with the answer that he *was* an ape." It is not sufficient to know that *Homo sapiens sapiens* evolved from a certain kind of primate. A more important question is what he would like or should try to become.

A mysterious aspect of human life is that throughout history social groups have abandoned their ancestral habits under the influence of abstract ideas; they have repeatedly rejected their traditions and modified their ways for the sake of an ideal. Equally mysterious is the fact that throughout prehistory and history almost all people have created monumental structures and have engaged in spectacular ceremonies of a purely symbolic nature to which they have devoted more efforts and resources than to the satisfaction of their material needs. The social life of the human species thus seems to be inspired by dreams which make it prefer change to status quo, symbolic creations to the production of creature comforts.

There exists a form of *joie de vivre* which expresses the very pleasure of being alive—a purely organic satisfaction which animals share with the human species. But there is another form of happiness, which seems peculiar to human beings. It originates from their deep awareness that their personal life is the realization of their dreams and their collective life a creative enterprise which gives concrete forms to the dreams of humankind.

Envoi

The emergence and evolution of humanity result from consecutive social adaptations to crises caused by natural disturbances and by willful changes.

Over many millennia, human beings have rarely if ever submitted passively to the influences of the environment. They have searched for the conditions they wanted and created them if these did not exist. They have thus been able constantly to renew their civilizations through adaptive social responses to self-selected environments.

Crises are practically always a source of enrichment and of renewal because they encourage the search for new solutions. These solutions cannot come from a transformation of human nature, because it is not possible to change the genetic endowment of the human species. But they can come from the manipulation of social structures because these affect the quality of behavior and of the environment, and therefore the quality of life.

Civilizations are mortal, but they can be revived and transformed by human imagination, fantasy, and will. It is through the death and resurrection of civilizations that humankind consecutively takes the many forms of which it is capable. Chance—or providence—provides the conditions and materials out of which civilizations can be reborn and renew themselves, but it is the human mind that selects among these options and that organizes the raw stuff of nature into humanized forms. We are human to the extent that we are able and willing to make the choices that enable us to transcend genetic and environmental determinism, and thus to participate in the continuous process of self-creation which seems to be the task and the reward of humankind.

Notes
Index

Notes

The following notes include primarily recent publications; they do not constitute complete documentation of the text. Notes beginning with "See" are suggestions for additional reading on the topic under discussion; others document quotations or ideas from specific sources.

I / Stability and Adaptability of Humankind

[1] See Gordon Willey, *An Introduction to American Archaeology. Vol. I. North and Middle America* (Englewood Cliffs, N.J.: Prentice-Hall, 1966); Kenneth Macgowan and Joseph A. Hester, Jr., *Early Man in the World* (New York: The Natural History Library, 1962); John E. Pfeiffer, *The Emergence of Man* (New York: Harper & Row, 1969); K.W. Butzer, *Environment and Archaeology* (Chicago: Aldine Publishing Company, 1964); Bernard Campbell, *Human Evolution* (Chicago: Aldine Publishing Company, 1967); and the series The Emergence of Man (New York: Time-Life Books): Maitland A. Edey, *The Missing Link* (1972); Edmund White and Dale Brown, *The First Men* (1973); George Constable, *The Neanderthals* (1973); Tom Prideaux, *Cro-Magnon Man* (1973); Robert Claiborne, *The First Americans* (1973); Dora Jane Hamblin, *The First Cities* (1973).

[2] J.A. Crow, *The Epic of Latin America* (Garden City, N.Y.: Doubleday, 1946).

[3] Theodora Kroeber, *Ishi in Two Worlds: A Biography of the Last Wild Indian in North America* (Berkeley, Calif.: University of California Press, 1961).

[4] Albert Maori Kiki, *Ten Thousand Years in a Lifetime* (New York: Praeger, 1969).

[5] Edward O. Wilson, "Competitive and Aggressive Behavior" in J.F. Eisenberg and W.S. Dillon (eds.), *Man and Beast: Comparative Social Behavior* (Washington, D.C.: Smithsonian Institution Press, 1971), p. 206. For a general discussion, see the review of *Man and Beast* by Jerram L. Brown, "The Ethology of Homo Sapiens," *Science* 174 (1971), 1013.

[6] A.C. Allison, "Protection Afforded by Sickle Cell Trait against Subtertian Malarial Infection," *British Medical Journal* 1 (1954), 290–292.

[7] Frederick Hulse, "Adaptation, Selection, and Plasticity," in G.W. Lasker (ed.), *The Processes of Ongoing Human Evolution* (Detroit: Wayne State University Press, 1960), pp. 74–75.

[8] See René Dubos, *So Human an Animal* (New York: Charles Scribner's Sons, 1968), and *A God Within* (New York: Charles Scribner's Sons, 1972).

[9] Peter Farb, *Word Play: What Happens When People Talk* (New York: Knopf, 1974).

[10] François Bordes, *A Tale of Two Caves* (New York: Harper, 1973).

[11] L.S. Cressman, *The Sandal and the Cave* (Portland, Ore.: Beaver Books, 1964).

II / Choosing to Be Human

[1] See Robert Ardrey, *African Genesis* (New York:

Atheneum, 1961), *The Territorial Imperative* (New York: Atheneum, 1966), and *The Social Contract* (New York: Atheneum, 1970); Macfarlane Burnet, *Dominant Mammal* (Melbourne, Australia: Heinemann, 1970); Konrad Lorenz, *On Aggression* (New York: Harcourt, Brace, and World, 1966); Desmond Morris, *The Naked Ape* (New York: McGraw-Hill, 1967), and *The Human Zoo* (New York: McGraw-Hill, 1969); Lionel Tiger and Robin Fox, *The Imperial Animal* (New York: Holt, Rinehart and Winston, 1971). For an opposite opinion, presenting the potentialities for good of the human species, see Alexander Alland, *The Human Imperative* (New York: Columbia University Press, 1972).

[2] Charles Hartshorne, *Born to Sing* (Bloomington, Ind.: Indiana University Press, 1973).

[3] George B. Schaller, *The Serengeti Lion: A Study of Predator-Prey Relations* (Chicago: University of Chicago Press, 1972). See also review by E.O. Wilson, "The Natural History of Lions," *Science* 179 (1973), 466–467; and letter by S.L. Washburn, *Psychology Today* 7 (1973), 4,6.

[4] Ralph Solecki, *Shanidar* (New York: Knopf, 1971).

[5] Quoted in John E. Pfeiffer, *The Emergence of Man* (New York: Harper & Row, 1969), p. 163.

[6] L.S. Cressman, *The Sandal and the Cave* (Portland, Ore.: Beaver Books, 1964).

[7] Solecki, *Shanidar*.

[8] Paolo Graziosi, *Palaeolithic Art* (New York: McGraw-Hill, 1960); André Leroi-Gourhan, *Treasures of Prehistoric Art,* trans. Norbert Guterman (New York: Abrams, 1968).

[9] Alexander Marshack, *The Roots of Civilization*

(New York: McGraw-Hill, 1970), and "Upper Paleolithic Notation and Symbol," *Science* 178 (1972), 817–828.

[10] Barbara Ward, *Nationalism and Ideology* (New York: Norton, 1966), p. 21.

[11] See René Dubos, *A God Within,* (1972), and *So Human an Animal* (1968) (New York: Charles Scribner's Sons).

[12] Edward T. Hall, *The Silent Language* (New York: Doubleday, 1959); and Robert Sommer, *Personal Space: The Behavioral Basis of Design* (Englewood Cliffs, N.J.: Prentice-Hall, 1969).

[13] Feodor Dostoevski, *Notes from Underground,* in *The Short Novels of Dostoevsky,* trans. Constance Garnett (New York: Dial Press, 1945). pp. 145, 149.

[14] Maxim Gorki, quoted in Trigant Burrow, *Preconscious Foundations of Human Experience* (New York: Basic Books, 1964), p. 114.

[15] Norman Mailer, *Advertisements for Myself* (New York: Signet, 1959), p. 304.

[16] Marcel Mauss, "Essai sur le don," *Sociologie et anthopologie* (Paris: Presses Universitaires de France, 1968); see also Jean Cazeneuve, *Sociologie de Marcel Mauss* (Paris: Presses Universitaires de France, 1968).

[17] Bent Jensen, "Human Reciprocity: An Arctic Exemplification," *American Journal of Orthopsychiatry,* April 1973.

[18] Wilton S. Dillon, *Gifts and Nations* (Paris: Mouton, 1968).

[19] See note 1, this chapter.

[20] E.A. Speiser (ed.), *The World History of the Jewish People,* vol. I. *At the Dawn of Civilization* (New Brunswick, N.J.: Rutgers University Press, 1964), p. 150.

²¹ Colin Turnbull, *The Mountain People* (New York: Simon & Schuster, 1972).

²² Quoted in Ernst Cassirer *et al.*, *The Renaissance Philosophy of Man* (Chicago: University of Chicago Press, 1948), pp. 224–225.

III / The Past in the Present

¹ Sigfried Giedion, *Space, Time and Architecture* (Cambridge, Mass.: Harvard University Press, 1967), pp. 417, 519.

² James Marston Fitch and Daniel P. Branch, "Primitive Architecture and Climate," *Scientific American* 203 (1960), 134–144; and Bernard Rudofsky, *Architecture without Architects* (Garden City, N.Y.: Doubleday, 1969).

³ Constantinos A. Doxiadis, *Architectural Space in Ancient Greece* (Cambridge, Mass.: Massachusetts Institute of Technology Press, 1972); Vincent Scully, *The Earth, the Temple and the Gods* (New York: Praeger, 1962).

⁴ Quoted in Constantine A. Doxiadis, *City for Human Development,* Ace Publication Series Research Report No. 12 (Athens: Athens Center of Ekistics, 1971), pp. 40, 64.

⁵ C.P. Cavafy, "The City," in *Four Greek Poets,* trans. Edmund Keeley and Philip Sherrard (London, Penguin, 1970), p. 13.

⁶ See G. Bell and J. Tyrwhitt, *Human Identity in the Urban Environment* (Baltimore, Md.: Pelican, 1973), p. 312; A.V.S. DeReuck and Ruth Porter (eds.), *Transcultural Psychiatry* (Boston: Little, Brown, 1966); William Michelson, *Man and His Urban Environment: A Sociological Approach* (Reading, Mass.: Addison-Wesley Publishing Company, 1970), p. 167.

[7] Richard B. Lee and Irven DeVore (eds.), *Man the Hunter* (Chicago: Aldine Publishing Company, 1968).

[8] W.F. Pratt, "Anabaptist Explosion; Adaptation of Pockets of High Fertility in the United States," *Natural History* 78 (1969), 8–10.

[9] Quoted in John E. Pfeiffer, *The Emergence of Man* (New York: Harper & Row, 1969), p. 334.

[10] Joyce Maynard, *Looking Backward* (Garden City, N.Y.: Doubleday, 1973).

[11] Livy, *The History of Rome* (Loeb Classical Library, Vol. 54, 1924). Also quoted in E.A. Gutkind, *International History of City Development*. Vol. 4, *Urban Development in the South: Italy and Greece* (Glencoe, Ill.: The Free Press, 1964), p. 28.

[12] See, for examples of people returning to devastated areas, Dorothy B. Vitaliano, *Legends of the Earth: Their Geologic Origins* (Bloomington, Ind.: Indiana University Press, 1973); J. Wohlwill and D.H. Carson, *Environment and the Social Sciences* (Washington, D.C.: American Psychological Association, 1972).

[13] John Rewald, *The History of Impressionism* (New York: New York Graphic Society, 1973).

[14] See for example K. Elliott (ed.), *The Family and Its Future* (London: J.A. Churchill, 1970); Jonathan L. Freedman, Simon Klevansky, and Paul Ehrlich, "The Effect of Crowding on Human Task Performance," *Journal of Applied Social Psychology* 1 (1971), 7–25; Jonathan L. Freedman, Alan S. Levy, Roberta Welte Buchanan, and Judy Price, "Crowding and Human Aggressiveness," *Journal of Experimental Social Psychology* 8 (1972), 528–548; Wohlwill and Carson, *Environment and the Social Sciences*.

[15] Patricia Draper, "Crowding among Hunter-

Gatherers: The !Kung Bushmen,'' *Science* 182 (1973), 303.

IV / At Home on Earth

[1] Alvin Toffler, *Future Shock* (New York: Random House, 1970).

[2] Quotations from Sumerian and Egyptian sources are from E.A. Speiser (ed.), *The World History of the Jewish People,* vol. I. (New Brunswick, N.J.: Rutgers University Press, 1964), pp. 150, 297, 299, 300.

[3] Louis Le Roy, *De la Vicissitude ou variété des choses en l'univers, et concurrence des armes et des lettres par les premières et plus illustres nations du monde, depuis le temps ou a commencé la civilité & mémoire humaine iusques à present* (Paris: Pierre L'Huillier, 1575); see also Werner L. Gundersheimer, *The Life and Works of Louis Le Roy* (Geneva: Librairie Droz, 1966).

[4] Henry Adams, *The Education of Henry Adams* (Boston: Houghton Mifflin, 1906).

[5] James Morris, ''The Final Solution Down Under,'' *Horizon* 14 (Winter, 1972).

[6] See note 1, chapter II; and also Carl O. Sauer, *Land and Life* (Berkeley, Calif.: University of California Press, 1963), and *Agricultural Origins and Dispersals: The Domestication of Animals and Foodstuffs* (Cambridge, Mass.: Massachusetts Institute of Technology Press, 1969).

[7] Jack Kerouac, *On the Road* (New York: The Viking Press, 1957).

[8] Nikos Kazantzakis, *The Odyssey: A Modern Sequel,* trans. Kimon Friar (New York: Simon & Schuster, 1958).

[9] Frederick Jackson Turner, *The Frontier in American History* (New York: Holt, 1920). See also R.A. Billington, *F. J. Turner* (New York: Oxford University Press, 1973); Richard Hofstadter, *The Progressive Historians* (New York: Knopf, 1968).

[10] Wilbur Zelinsky, *The Cultural Geography of the United States* (Englewood Cliffs, N.J.: Prentice-Hall, 1973), p. 83.

[11] Vannevar Bush, *Science, the Endless Frontier* (Washington, D.C.: National Science Foundation, 1945; reprinted 1960).

[12] D.H. Meadows *et al., The Limits to Growth* (New York: Universe Books, 1972).

[13] Edward Goldsmith *et al., Blueprint for Survival* (Boston: Houghton Mifflin, 1972).

[14] See Dennis Gabor, *Innovations: Scientific, Technological and Social* (New York: Oxford University Press, 1971); *Inventing the Future* (New York: Knopf, 1964); *The Mature Society* (New York: Praeger, 1972).

[15] See Paul Shepard, *The Tender Carnivore and the Sacred Game* (New York: Charles Scribner's Sons, 1973), chapter 6; Raymond Williams, *The Country and The City* (New York: Oxford University Press, 1973).

[16] José Ortega y Gasset, *Meditations on Hunting,* trans. Howard Wescott (New York: Charles Scribner's Sons, 1972), pp. 141, 142.

[17] Paul Valéry, "Vivant," *Cahiers du Sud,* June 1972, p. 354, quoted from Paul Valéry, *Variété I.*

V / On the Pleasures of Being Human

[1] Aleksandr I. Solzhenitsyn, "Art—For Man's Sake," trans. Thomas Whitney (The Nobel Foundation, 1972).

[2] Edward H. Spicer, "Persistent Cultural Sys-

tems," *Science* 172 (1971), 795–800.

[3] Quoted in *Review of National Literatures: Russia—The Spirit of Nationalism* 3 (1973), 140.

[4] See W.C. Bark, *Origins of the Medieval World* (Garden City, N.Y.: Doubleday, 1962; G.C. Coulton, *The Medieval Scene* (Cambridge: Cambridge University Press, 1930); Denys Hay, *The Medieval Centuries* (London: Methuen, 1953); Lynn White, Jr., "What Accelerated Technological Progress in the Western Middle Ages," in Francis Oakley and Daniel O'Connor (eds.), *Creation: The Impact of an Idea* (New York: Charles Scribner's Sons, 1969).

[5] Michel de Montaigne, *The Complete Essays,* translated by D. Frame (Palo Alto, Calif.: Stanford University Press, 1958), p. 615.

[6] René Dubos, *A God Within* (New York: Charles Scribner's Sons, 1972).

[7] Dora Jane Hamblin, *The First Cities* (New York: Time-Life Books, 1973).

[8] Claude Lévi-Strauss, *Race et Histoire* (Paris: Denoël Gonthier, 1961).

[9] D.H. Meadows *et al., The Limits to Growth* (New York: Universe Books, 1972).

[10] Dennis Gabor, *Inventing the Future* (New York: Knopf, 1964).

[11] Denise de Sonneville-Bordes, "Upper Paleolithic Cultures in Western Europe," *Science* 142 (1963), 355.

[12] Jo Ann Levine, "Barbara Ward," *Christian Science Monitor,* November 30, 1973, p. 21.

[13] Albert J. Fritsch, *A Theology of the Earth* (Washington, D.C.: CLB Publishers, 1972).

[14] Robert Graves, *The Word Báraka.* The Blashfield Address (New York: American Academy of Arts and Letters, 1962).

Index

222

About the Author

RENÉ DUBOS, professor emeritus at The Rockefeller University in New York City, is a microbiologist and experimental pathologist who more than a quarter of a century ago was the first to demonstrate the feasibility of obtaining germ-fighting drugs from microbes. For his scientific contributions, Dr. Dubos has received many awards; most recently he was the recipient of the first Institut de la Vie prize for his work devoted to environmental problems. He is well known as an author and lecturer as well as a scientific investigator; his books include *So Human an Animal* (1969 Pulitzer Prize winner), *A God Within, Reason Awake! Science for Man, Only One Earth* (with Barbara Ward), *Louis Pasteur—Free Lance of Science, The Torch of Life, The Unseen World, The Dreams of Reason, The White Plague* (with Jean Dubos), *The Mirage of Health,* and *Man Adapting.*